# DNA Testing

# DNA Testing

by

**B.S. Bedi**

**MOHIT PUBLICATIONS**

**New Delhi 110 002**

ISBN 81-7445-364-4

*Published by :*

**MOHIT PUBLICATIONS**

4675/21, Ansari Road, Darya Ganj
New Delhi 110 002
Phones : 23255141, 23284136

PRINTED IN INDIA

---

Published by P.K. Goel for Mohit Publications, New Delhi-110002,
Typeset by SCS, Printed at

# PREFACE

DNA Testing is a technique used to distinguish between individuals of the same species using only samples of their DNA. Its invention by Sir Alec Jeffreys at the University of Leicester was announced in 1985. Two humans will have the vast majority of their DNA sequence in common. Genetic fingerprinting exploits highly variable repeating sequences called microsatellites. Two unrelated humans will be likely to have different numbers of microsatellites at a given locus. By using PCR to detect the number of repeats at several loci, it is possible to establish a match that is extremely unlikely to have arisen by coincidence, except in the case of identical twins, who will have identical genetic profiles.

Genetic fingerprinting is used in forensic science, to match suspects to samples of blood, hair, saliva or semen. It has also led to several exonerations of formerly convicted suspects. It is also used in such applications as studying populations of wild animals, paternity testing, identifying dead bodies, and establishing the province or composition of foods. It has also been used to generate hypotheses on the pattern of the human diaspora in prehistoric times.

Testing is subject to the legal code of the jurisdiction in which it is performed. Usually the testing is voluntary, but it can be made compulsory by such instruments as a search warrant or court order. Several jurisdictions have also begun to assemble databases containing DNA information of convicts. The United Kingdom currently has the most extensive DNA database in the world, with well over 2 million records as of 2005: The National DNA Database (NDNAD). The size of this database, and its rate of growth, is giving concern to civil liberties groups in the UK, where police have wide-ranging powers to take samples and retain them even in the event of acquittal.

DNA fingerprinting begins by extracting DNA from the cells in a sample of blood, saliva, semen, or other appropriate fluid or tissue. Reference samples are often collected using a buccal swab. When DNA fingerprinting first began, restriction fragment length polymorphism (RFLP) analysis was used, though it has been almost completely replaced with newer techniques. RFLP analysis is performed by using a restriction enzyme to cut the DNA into fragments which are separated into bands during agarose gel electrophoresis. Next, the bands of DNA are transferred via a technique called Southern blotting from the agarose gel to a nylon membrane. This is treated with a radioactively-labelled DNA probe which binds to certain specific DNA sequences on the membrane. The excess DNA probe is then washed off. An X-ray film placed next to the nylon membrane detects the radioactive pattern. This film is then developed to make a visible pattern of bands called a DNA fingerprint. By using multiple probes targeting various polymorphisms in successive X-ray images, a fairly high degree of discrimination was possible. The primary drawback of RFLP is that the exact sizes of the bands are unknown and comparison to a molecular weight ladder is done in a purely qualitative manner. This made typing degraded samples such as those from evidence that had been exposed to the elements fairly difficult.

**B.S. Bedi**

# CONTENTS


*Preface*

1. DNA Testing: An Introduction 1
2. Methods and Strategies for DNA Sequences 11
3. Trouble with DNA 37
4. DNA Matching 62
5. Power of DNA 85
6. DNA Computing 96
7. DNA Databases for 21st Century 105
8. Forensic DNA Kits 163
9. Genetic Discrimination and the Workplace 187

*Bibliography* 225

*Index* 229

# 1

# DNA Testing: An Introduction

## Overview

DNA Testing is the genetic blueprint that determines a person's biological characteristics. DNA is located in the cell of the human body. Upon conception, a child inherits one half of its DNA from its mother and one half from its father. This unique combination of DNA will match that of the biological parents of the child. For the test to take place, cheek cells are removed with a swab from the mouth of the mother, the child and the alleged father. In the laboratory, these cells are removed from the swab and tested. When a child's DNA does not match that of the alleged father, he is excluded 100% as the biological father of the child. When the child's DNA does match that of the alleged father, this means that the alleged father "could be" the father of the child, but it does not prove that he "is" the only possible father, since related individuals, such as brothers, can share DNA characteristics. However, a DNA test can prove a probability of 99% or greater for paternity.

DNA testing and DNA profiling are techniques used to distinguish between individuals of the same species using only samples of their DNA. Its invention by Sir Alec Jeffreys at the University of Leicester was announced in 1985. Two humans will have the vast majority of their DNA sequence in common. Genetic fingerprinting exploits highly variable repeating sequences called microsatellites. Two unrelated humans will be likely to have different numbers of microsatellites at a given locus. By using PCR to detect the number of repeats at several loci, it is possible to establish a match that is extremely unlikely to have arisen by coincidence, except in the case of identical twins, who will have identical genetic profiles.

Genetic fingerprinting is used in forensic science, to match suspects to samples of blood, hair, saliva or semen. It has also led to several exonerations of formerly convicted suspects. It is also used in such applications as studying populations of wild animals, paternity testing, identifying dead bodies, and establishing the province or composition of foods. It has also been used to generate hypotheses on the pattern of the human diaspora in prehistoric times.

Testing is subject to the legal code of the jurisdiction in which it is performed. Usually the testing is voluntary, but it can be made compulsory by such instruments as a search warrant or court order. Several jurisdictions have also begun to assemble databases containing DNA information of convicts.

The United Kingdom currently has the most extensive DNA database in the world, with well over 2 million records as of 2005: The National DNA Database (NDNAD). The size of this database, and its rate of growth, is giving concern to civil liberties groups in the UK, where police have wide-ranging powers to take samples and retain them even in the event of acquittal.

## DNA Fingerprinting Methods

DNA fingerprinting begins by extracting DNA from the cells in a sample of blood, saliva, semen, or other appropriate fluid or tissue. Reference samples are often collected using a buccal swab.

When DNA fingerprinting first began, restriction fragment length polymorphism (RFLP) analysis was used, though it has been almost completely replaced with newer techniques. RFLP analysis is performed by using a restriction enzyme to cut the DNA into fragments which are separated into bands during agarose gel electrophoresis. Next, the bands of DNA are transferred via a technique called Southern blotting from the agarose gel to a nylon membrane. This is treated with a radioactively-labelled DNA probe which binds to certain specific DNA sequences on the membrane.

The excess DNA probe is then washed off. An X-ray film placed next to the nylon membrane detects the radioactive pattern. This film is then developed to make a visible pattern of bands called a DNA fingerprint. By using multiple probes targeting various polymorphisms in successive X-ray images, a fairly high degree of discrimination was possible. The primary drawback of RFLP is that the exact sizes of the bands are unknown and comparison to a molecular weight ladder is done in a purely qualitative manner. Many labs developed policies that described what they considered a unique band, but it was not standardized and led to DNA fingerprinting coming under harsh attack in People v. RFLP was a very time consuming method which required relatively high quantity of good quality DNA to be used (such as a dime sized blood drop). This made typing degraded samples such as those from evidence that had been exposed to the elements fairly difficult.

With the invention of polymerase chain reaction (PCR), DNA fingerprinting took huge strides forward in both discriminating power and ability to recover information from very small starting samples. PCR involves the amplification of specific regions of DNA using a cycling of temperature and a thermostable polymerase enzyme along with sequence specific primers of DNA. Commercial kits that used single nucleotide polymorphisms (SNPs) for discrimination became available. These kits use PCR to amplify specific regions with known variations and hybridize them to probes anchored on cards, which results in a colored spot corresponding to the particular sequence variation.

One of the primary complaints against RFLP was that it was slow and required large quantities of DNA to be used. This led to the development of PCR-based methods which required smaller amounts of DNA that could also be more degraded than those used in RFLP analysis. Systems such as the HLA-DQ alpha reverse dot blot strips grew to be very popular due to their ease of use and the speed with which a result could be obtained, however they were not as discriminating as RFLP. It was also difficult to determine a DNA profile for mixed samples, such as a vaginal swab from a sexual assault victim.

Another technique, AmpFLP, or amplified fragment length polymorphism was also put into practice during the early 1990's. This technique was also faster than RFLP analysis and used PCR to amplify DNA samples. It relied on variable number tandem repeat (VNTR) polymorphisms to distinguish various alleles, which were separated on a polyacrylamide gel using an allelic ladder (as opposed to a molecular weight ladder). Bands could be visualized by silver staining the gel. One popular locus for fingerprinting was the D1S80 locus. As with all PCR based methods, highly degraded DNA or very small amounts of DNA may cause allelic dropout (causing a mistake in thinking a heterozygote is a homozygote) or other stochastic effects. In addition, because the analysis is done on a gel, very high number repeats may bunch together at the top of the gel, making it difficult to resolve. AmpFLP analysis can be highly automated, and allows for easy creation of phylogenetic trees based on comparing individual samples of DNA. Due to its relatively low cost and ease of set-up and operation, AmpFLP remains popular in lower income countries.

The most prevalent method of DNA fingerprinting used today is based on PCR and uses short tandem repeats (STR). This method uses highly polymorphic regions that have short repeated sequences of DNA (the most common is 4 bases repeated, but there are lengths in use, including 3 and 5 bases). Because different people have different numbers of repeat units, these regions of DNA can be used to discriminate between individuals. These STR loci (locations) are targeted with sequence-specific primers and are amplified using PCR. The DNA fragments that result are then separated and detected using electrophoresis. There are two common methods of separation and detection, capillary electrophoresis (CE) and gel electrophoresis.

The polymorphisms displayed at each STR region are by themselves very common, typically each polymorphism will be shared by around 5 - 20% of individuals. When looking at multiple loci, it is the unique combinations of these polymorphisms to an individual that makes this method discriminating as an identification tool. The more STR regions that are tested in an individual the more discriminating the test becomes.

From country to country different STR based DNA profiling systems are in use. In North America CODIS is prevalent, while in the UK the SGM+ system, which is compatible with The National DNA Database is in use. Whichever system is used, many of the STR regions under test are the same. These DNA profiling systems are based around multiplex reactions, whereby many STR regions will be under test at the same time.

The Applied Biosystems 3130xl Genetic Analyzer is a popular instrument that can perform capillary electrophoresis to type DNA. Capillary electrophoresis works by electrokinetically (movement through the application of an electric field) injecting the DNA fragments into a thin glass tube (the capillary) filled with polymer. The DNA is pulled through the tube by the application of an electric field, separating the fragments such that the smaller fragments travel faster through the capillary. The fragments are then detected using fluorescent dyes that were attached to the primers used in PCR. This allows multiple fragments to be amplified and run simultaneously, something known as multiplexing. Sizes are assigned using labeled DNA size standards that are added to each sample, and the number of repeats are determined by comparing the size to an allelic ladder, a sample that contains all of the common possible repeat sizes. Although this method is expensive, larger capacity machines with higher throughput are being used to lower the cost/sample and reduce backlogs that exist in many government crime facilities.

Gel electrophoresis acts using similar principles as CE, but instead of using a capillary, a large polyacrylamide gel is used to separate the DNA fragments. An electric field is applied, as in CE, but instead of running all of the samples by a detector, the smallest fragments are run close to the bottom of the gel and the entire gel is scanned into a computer. This produces an image showing all of the bands corresponding to different repeat sizes and the allelic ladder. This approach does not require the use of size standards, since the allelic ladder is run alongside the samples and serves this purpose. Visualization can either be through the use of fluorescently tagged dyes in the primers or by silver staining the gel prior to scanning. Although it is cost effective and can be rather high throughput, silver staining kits for STRs are being discontinued. In addition, many labs are phasing out gels in favor of CE as the cost of machines becomes more manageable.

The true power of STRs is in its statistical power of discrimination. In the U.S.A., there are 13 loci (DNA locations) that are currently used for discrimination. Because these loci are independently assorted (having a certain number of repeats at one locus doesn't change the likelihood of having any number of repeats at any other locus), the power rule of statistics can be applied. This means that if someone has the DNA type of ABC, where the three loci were independent, we can say that the probability of having that DNA type is the probability of having type A times the probability of having type B times the probability of having type C. This has resulted in the ability to generate match probabilities of 1 in a quintillion (1 with 18 zeros after it) or more. Recent innovations have included the creation of primers targeting polymorphic regions on the Y-chromosome (Y-STR), which allows resolution of multiple male profiles, or cases in which a differential extraction is not possible. Y-chromosomes are paternally inherited, so Y-STR analysis can help in the identification of paternally related males. Y-STR analysis was performed in the Sally Hemings controversy to determine if Thomas Jefferson had sired a son with one of his slaves.

For highly degraded samples, it is sometimes impossible to get a complete profile of the 13 CODIS STRs. In these situations, mitochondrial DNA (mtDNA) is sometimes typed due to there being many copies of mtDNA in a cell, while there may only be 1-2 copies of the nuclear DNA. Forensic scientists amplify the HV1 and HV2 regions of the mtDNA, then sequence each region and compare single nucleotide differences to a reference. Because mtDNA is maternally inherited, directly linked maternal relatives can be used as match references, such as one's maternal grandmother's sister's son. A difference of two more nucleotides is generally considered to be an exclusion. Heteroplasmy and poly-C differences may throw off straight sequence comparisons, so some expertise on the part of the analyst is required. mtDNA is useful in determining unclear identities, such as those of missing persons when a maternally linked relative can be found. mtDNA testing was used in determining that Anna Anderson was not the Russian princess she had claimed to be, Anastasia Romanov.

## Considerations when Evaluating DNA Evidence

In the early days of the use of genetic fingerprinting as criminal evidence, juries were often swayed by spurious statistical arguments by defense lawyers along these lines: given a match that had a 1 in 5 million probability of occurring by chance, the lawyer would argue that this meant that in a country of say 60 million people there were

12 people who would also match the profile. This was then translated to a 1 in 12 chance of the suspect being the guilty one. This argument is not sound unless the suspect was drawn at random from the population of the country. In fact, a jury should consider how likely it is that an individual matching the genetic profile would also have been a suspect in the case for other reasons. Another spurious statistical argument is based on the false assumption that a 1 in 5 million probability of a match automatically translates into a 1 in 5 million probability of innocence and is known as the prosecutor's fallacy.

When using RFLP, the theoretical risk of a coincidental match is 1 in 100 billion (100,000,000,000). However, the rate of laboratory error is almost certainly higher than this, and often actual laboratory procedures do not reflect the theory under which the coincidence probabilities were computed. For example, the coincidence probabilities may be calculated based on the probabilities that markers in two samples have bands in precisely the same location, but a laboratory worker may conclude that similar – but not precisely identical -- band patterns result from identical genetic samples with some imperfection in the agarose gel. However, in this case, the laboratory worker increases the coincidence risk by expanding the criteria for declaring a match. Recent studies have quoted relatively high error rates which may be cause for concern [1]. Because of this, arbitrary ceilings were put on match probabilities used in RFLP analysis than the theoretically computed ones. Today, RFLP has become widely disused due to these difficulties in interpretation.

STRs do not suffer from such subjectivity and provide much better powers of discrimination, for unrelated individuals (of the order of 1 in 10^29 if using a full profile) It should be noted that figures of this magnitude are not considered to be statistically suportable by scientists in the UK, for unrelated individuals with full matching DNA profiles a match probability of 1 in a billion (one thousand million) is considered statistically supportable (Since 1998 the DNA profiling system supported by The National DNA Database in the UK is the SGM+ DNA profiling system which includes 10 STR regions and a sex indicating test, this test updated the SGM DNA profiling system on which the National DNA Database was founded in 1995. The SGM system included 6 out of the 10 STR regions used in the SGM+ system and the same sex indicating test, however the discriminating power of the SGM system was only considered to be supportable at 1 in a million) . However, with any DNA technique, the cautious juror should not convict on genetic fingerprint evidence alone if other factors raise doubt. Contamination with other evidence (secondary transfer) is a key source of incorrect DNA profiles and raising doubts as to whether a sample has been adulterated is a favorite defense technique. More rarely, Chimerism is one such instance where the lack of a genetic match may unfairly exclude a suspect.

When evaluating a DNA match, the following questions should be asked:

- Could it be an accidental random match?
- If not, could the DNA sample have been planted?
- If not, did the accused leave the DNA sample at the exact time of the crime?
- If yes, does that mean that the accused is guilty of the crime?

## Fake DNA Evidence

The value of DNA evidence has to be seen in light of recent cases where criminals planted fake DNA samples at crime scenes. In one case, a criminal even planted fake DNA evidence in his own body: Dr. John Schneeberger of Canada raped one of his sedated patients in 1992 and left semen on her underwear. Police drew Schneeberger's blood and compared its DNA against the crime scene semen DNA on three occasions, never showing a match. It turned out that he had surgically inserted a Penrose drain into his arm and filled it with foreign blood and anticoagulants.

## Cases

In 1987, British baker Colin Pitchfork was the first criminal caught using DNA fingerprinting.

In 1987, Florida rapist Tommie Lee Andrews was the first person to be convicted as a result of DNA evidence, for raping a woman during a burglary; he was convicted on 6 November 1987 and sentenced to 22 years in prison. He later had the sample retested because of concerns over the reliability of early genetic fingerprinting techniques.

In 1989, Chicago man Gary Dotson was the first person whose conviction was overturned using DNA evidence.

In 1991, Allan Legere was the first Canadian to be convicted as a result of DNA evidence, for four murders he had committed while an escaped prisoner in 1989. During his trial, his defense argued that the relatively shallow gene pool of the region could lead to false positives.

In 1992, DNA evidence was used to prove that Nazi doctor Josef Mengele was buried in Brazil under the name Wolfgang Gerhard.

The science was made famous in the United States in 1994 when prosecutors heavily relied on - and through expert witnesses exhaustively presented and explained - DNA evidence allegedly linking O. J. Simpson to a double murder. The case also brought to light the laboratory difficulties and handling procedure mishaps which can cause such evidence to be significantly doubted.

In 1994, RCMP detectives successfully tested hairs from a cat known as, Snowball, and used the test to link a man to the murder of his wife, thus marking for the first time in forensic history the use of non-human DNA to identify a criminal.

In 1998, Dr. Richard J. Schmidt was convicted of attempted second-degree murder when it was shown that there was an indisputable link between the viral DNA of the HIV virus he had been accused of injecting in his girlfriend and one of his patients with full-blow AIDS. This was the first time viral DNA fingerprinting had been used as evidence in a criminal trial.

In 2003, Welshman Jeffrey Gafoor was convicted of the 1988 murder of Lynette White, when crime scene evidence collected 12 years earlier was re-examined using STR techniques, resulting in a match with his nephew. This may be the first known example of the DNA of an innocent yet related individual being used to identify the actual criminal, via "familial searching".

In June of 2003, because of new DNA evidence, Dennis Halstead, John Kogut and John Restivo won a re-trial on their murder conviction. The three men had already served eighteen years of their thirty plus year sentences.

The trial of Robert Pickton is notable in that DNA evidence is being used primarily to identify the victims, and in many cases to prove their existence.

In March 2003, Josiah Sutton was released from prison after serving four years of a twelve year sentence for a sexual assault charge. Questionable DNA samples taken from Sutton were retested in the wake of the Houston Police Department's crime lab scandal of mishandling DNA evidence.

In December 2005, Robert Clark was proven innocent of a 1981 attack on an Atlanta woman after serving twenty four years in prison. Mr Clark is the 164th person in United States and the fifth in Georgia to be freed using post-conviction DNA testing.

# 2

# METHODS AND STRATEGIES FOR DNA SEQUENCES

## Introduction

Sequence data of DNA are becoming increasingly important within systematic and evolutionary biology. With this expanding emphasis, many investigators have rising expectations that sequence variation offers an almost unlimited potential to resolve phylogenetic relationships and thereby provide us with a highly robust, if not definitive, description of the pattern of life's history. Various reasons have been given in support of this expectation. One of the most common is that because DNA is the physical information system recording all inherited attributes of organismal history, direct comparison of its structure provides the most basic of all data for phylogcnetic reconstruction. Inasmuch as all historical change is encoded in DNA, it is reasoned, it makes sense to examine that change directly. A second argument is one of large numbers: we can obtain far more character-state data from DNA sequences than we can, say, from traditional morphological comparisons, and therefore, statistically, as additional data are examined, true patterns of phylogenetic relationships will eventually emerge. Furthermore, it is often claimed that because morphology is so readily subject to convergence when compared to sequence data, obtaining more and more of the latter represents our best hope for resolving the phylogenetic pattern.

Technological innovations are also fueling the rush to obtain sequence data. Comparative sequence data are much easier and less expensive to collect than they were even a year ago. Just as importantly, there has been a rapid growth in the amount of computer hardware and software that makes the analysis of comparative sequence data readily accessible to workers at all levels of expertise. In particular, the availability of tree-building algorithms for microcomputers has been a contributing factor to the marked increase in the numbers of papers that use DNA sequences to answer phylogenetic questions. At its simplest, therefore, it is relatively straightforward to collect comparative sequence data, analyze them with a given tree-building method, and publish the results.

Yet, it is clear from the literature that phylogenetic inference using sequence data is anything but simple. There is, for example, substantial disagreement over the best method to resolve relationships from sequence data. And even if agreement could be reached on a general method, it still is not clear how that method might be applied in order to obtain a phylogenetic hypothesis that is judged reliable, or best, by some criterion. These and other problems make phylogenetic analysis of DNA sequences extremely complex.

There are three main classes of methods of phylogenetic inference applied to sequence data. First, with distance analysis, a distance (similarity/dissimilarity) coefficient is computed for all pairwise comparisons of raw sequence data, or for data corrected for multiple substitutions, and the taxa are then clustered based on the values in the distance matrix. In general, the tree having the shortest overall branch length relative to that implied by the original data is chosen as the preferred tree. Second, the method of maximum likelihood uses the original sequence data and, working from a prior evolutionary model of nucleotide substitution, computes a likelihood score for each tree given the original data. That tree with the highest likelihood is the most preferred. The third method, cladistics or parsimony analysis, also uses the original sequence data, and from calculations of character-state changes (nucleotide substitutions) on alternative trees, the investigator chooses the one having the fewest character-state transformations, that is, the one for which homoplasies parallelisms and reversals are minimized.

There are several important reasons for preferring a parsimony approach to phylogenetic inference when using DNA sequences. First, parsimony methods are applied to the original sequence data rather than to some transformed distance metric, thus avoiding an inevitable loss of information. Second, while not entirely assumption free, parsimony methods are either less dependent upon assumptions about sequence evolution than are other methods; or they at least permit a more informed investigation of any assumptions that might be made. Third, algorithms that implement parsimony procedures are far more sophisticated than distance or maximum-likelihood alternatives and consequently allow a deeper analysis of the phylogenetic structure of one's data and of the dynamics of sequence evolution. All three of these reasons mean that parsimony provides the investigator with enhanced interpretability of the results as compared to other methods.

Nevertheless, as noted above, the application of parsimony procedures to sequence data is complex and has not yet received sufficient attention. This chapter addresses two of the more general problems involving the use of parsimony procedures in phylogenetic inference: given that there are physico-chemical/functional constraints on sequence evolution, especially in sequences coding for proteins or structural RNAs, how might parsimony be applied in order to infer phylogenetic relationships, and how might we judge the phylogenetic informativeness of sequence data? Our inquiry into these problems will be based on an analysis of mitochondrial DNA (mtDNA), but the findings should be applicable to other parts of the genome as well.

Although it is a simple enough procedure to obtain a global parsimony estimate of the phylogenetic structure inherent in any given data set, recognition that rates of transitions are, at least in vertebrate mtDNA, much higher than those for transversions, that there may be biases in base composition, and that silent substitutions in the third codon position are more common than those at first and second positions, have suggested to some investigators that parsimony procedures will have difficulties when used to resolve phylogenetic relationships, especially among more distantly related taxa. If homoplasy is too extensive, it is argued, obtaining a strongly corroborated phylogenetic tree might be difficult because even a minimum-length tree will have a high number of homoplastic changes. To the extent that this is true, it raises several important questions: How might we undertake parsimony analysis so as to reduce, or at least identify, the influence of homoplasy? Second, how do we identify those parts of the tree that seem to be less strongly supported and at the same time specify which components of the data are less informative phylogenetically? Finally, in any given study, how do we determine how much sequence must be obtained in order to reveal a reliable phylogenetic signal?

These questions speak to the notion of *reliability and informativeness*, terms that have been used in different ways in the literature. Some investigators speak of "reliability of methods," referring to that method which finds the best-fit tree for a given set of data. Others view reliability in terms of an "ideal" result: the ability of a method, or a set of data, to enable us to infer a tree that reflects the one true phylogeny. It is well to keep the distinction of these two views of reliability in mind, for often it is implied that once a best-fit tree is obtained, so too has an accurate estimate of the one true phylogeny. It is sometimes argued, moreover, that under a particular model of nucleotide sequence change, one or more methods will fail to be reliable; that is, it would be expected that the best fit tree generated by those methods will not be an accurate estimate of the true phylogeny. All of these views on reliability, however, are more arguments over methods, or perhaps of models of evolutionary change, inasmuch as the one true phylogeny can never be specified with certainty. Nor is it clear that a "realistic" evolutionary model can be developed without some prior hyothesis of relationships with which to investigate character-state change. Therefore, given only the data set of interest, it is difficult to say whether the best-fit tree produced by any method is or is not an accurate representation of the true phylogeny; it is simply the most economical explanation for the available data.

Ideally, "reliability" is best thought of as a parsimony problem and consequently should be judged in terms of congruence of data. "Informativeness," likewise, is also a parsimony problem. The two concepts are related. A given tree, calculated to be the most-parsimonious, or best-fit, for some set of data, can be taken to be a reliable estimate of the true phylogenetic history of the taxa if most-parsimonious trees for other, independent data are congruent with it. A given sample of sequence can be judged phylogenetically informative if it corroborates the most-parsimonious tree derived from other data. This reciprocal duality between the structure of data and what is inferred from them manifests the application of parsimony (and congruence analysis) to hypotheses at two different levels of inference: that of the data (character homology or hypotheses of synapomorphy), and that of the phylogenetic hypotheses themselves.

## The Data

This study uses a single data set, and therefore, informativeness and reliability will be assessed in relation to the "stability" seen in the cladistic signal across partitioned subsets of the data. That is, to the extent to which clades are corroborated by these different subsets, the data will be judged informative and the phylogenetic hypotheses themselves reliable. "Noise," on the other hand, will be judged to exist when the data do not corroborate a consistent (or stable) cladistic signal.

The sequence data analyzed in this study include an 898 base pair fragment of mtDNA that encompasses 458 bp of the 3' end of the NADHdehydrogenase subunit 4 (ND4 gene), 198 bp of three transfer RNA genes ($tRNA^{His}$, $tRNA^{Ser}$, and $tRNA^{Leu}$), and 242 bp of the 5' end of the ND5 gene. Comparative sequences were available for 12 primate taxa: human (*Homo sapiens*), common chimpanzee (*Pan troglodytes*), gorilla (*Gorilla gorilla*), orangutan (*Pongo pygmaeus*), gibbon (*Hylobates lar*), Japanese macaque (*Macaca fuscata*), rhesus macaque (*M. mulatta*), crab-eating macaque (*M. fascicularis*), Barbary macaque (*M. sylvanus*), squirrel monkey (*Saimiri sciureus*), Philippine tarsier (*Tarsius syrichta*), and ring-tailed lemur (*Lemur caua*).

## Methods of Analysis

All analyses were undertaken using the Phylogenetic Analysis Using Parsimony (PAUP) computer program of D. L. Swofford. Version 3.0 of this program for the Macintosh system is especially well suited for analysis of sequence data because it readily permits investigation of the effects of transitions versus transversions, the inclusion/exclusion of first, second, or third codon positions in coding sequences, and of nucleotide substitution parsimony of amino acid replacements. In addition, the program can compute consensus trees by different methods, perform bootstrap analyses, and calculate numerous measures of the information content for a given tree structure.

In order to investigate the effects of sample size of characters on the cladistic structure of the data, the 898 bp fragment was partitioned into four sequential subsets: the first 225 bp of the ND4 gene; a 458 bp segment including all of the ND4 gene that was sequenced; a 656 bp fragment including the 458 bp ND4 fragment plus the 198 bp of the three tRNAs; and the entire 898 bp fragment representing the ND4 fragment, tRNAs, and the 242 bp fragment of the ND5 gene. The effects of data set size were studied using this approach (rather than by random subsamples of different sizes) because systematists add sequence data sequentially and because it permitted an analysis of regions of the data set having different functional domains. In all analyses, the trees were rooted by designating the lemur as the out group. The results would have been the same if the tarsier had been designated the out group; therefore, this study is only investigating patterns of relationships within the clade consisting of the other primate species. A nonprimate outgroup was not used in order to make our results more comparable to those of Hayasaka.

Hayasaka proposed a phylogenetic hypothesis for the 12 primate taxa based upon substitution differences among those taxa and using the neighbor-joining method of Saitou and Nei. A single tree was obtained, which according to Hayasaka, was also obtained using the unweighted pair-group method of distance analysis and by the distance-Wagner method. Hayasaka concluded that "Because these three different methods give phylogenetic trees with the same topology, the phylogenetic relationships derived from these mtDNA sequence comparisons appear reliable." It is not the purpose of this chapter to speak to the issue of primate relationships *per se*; rather the goal is to examine influences on the phylogenetic structure of these mtDNA sequences, to explore the implications of that structure for assessing the "reliability" of sequence data in general, and to investigate how parsimony analysis might be effectively applied to sequence data.

Two series of analyses are used to examine and describe the extent to which the primate mtDNA data set is phylogenetically informative. The first series employs parsimony analysis on different subsets of the data with the purpose of revealing the cladistic structure of most-parsimonious and near most-parsimonious trees. The most-parsimonious tree, or collection of equally parsimonious trees, was found. Once its length was known, all trees within five steps of that length were retained and a majority-rule consensus tree computed (majority-rule consensus trees were adopted because they permit an analysis of the relationship between the amount of phylogenetic signal and data set size and because they are comparable to the results of bootstrap analyses as discussed below). The collection of trees having lengths within 1% of the minimum-length tree was also examined and will be mentioned briefly.

A second series of analyses used the same subsets of data and constructed a majority-rule bootstrap tree from 100 randomly sampled (with replacement) replications of each of the subsets. The bootstrap, in this case, is not intended as a test of statistical significance but as an additional means of describing and evaluating the structure of the phylogenetic signal inherent in the data.

## Results

A global parsimony analysis was undertaken using all nucleotide substitutions in each of the four subsets of the 898 bp fragment. This analysis was designed to answer two principle questions: Can a global parsimony analysis, in which transitions and transversions are equally (uniformly) weighted, reveal a stable phylogenetic signal? And, second, does the stability of that signal change as sample size is increased? Most-parsimonious trees are described but not figured. Primary attention is given to the set of trees within five steps of the most-parsimonious tree (or trees), but the collection of trees within 1% of the minimum-length tree is also examined.

1. **ND4:225 bp Fragment.** A parsimony analysis of the smallest data set of 225 characters yielded six equally parsimonious trees of 264 steps. These trees have only two topological ambiguities: the hominine trichotomy; and a trichotomy among the hominoids, macaques, and the squirrel monkey.

All other relationships were as postulated by Hayasaka. There are 95 trees within five steps of the shortest trees. The consensus tree exhibits relatively poor structure except within the macaques. There is no resolution within the great apes clade or among the basal lineages of the anthropoids.

2. **ND4:458 bp Fragment.** When the analysis is expanded to include all of the ND4 fragment, a single most parsimonious tree of 608 steps is found that is similar to the distance tree except for a sister-group relationship between the chimp and gorilla.

   In the 458 bp data set for ND4 there are 17 trees within five steps of the single most-parsimonious tree. Resolution within the hominoids has improved substantially, although the human-chimpanzee-gorilla trichotomy, while recognized in 88% of the trees, is not resolved internally. The hominoid relationships of the gibbon are still somewhat ambiguous, occurring in only 53% of the trees. The additional data do not change the resolution within the macaques, but do now place them closer to the hominoids than to the squirrel monkey in 65% of the trees.

3. **ND4 + tRNAs: 656 bp Fragment.** A parsimony analysis of the ND4 fragment and the three tRNA genes produced a single most-parsimonious tree of 766 steps identical to that of the preceding analysis in which the gorilla is joined to the chimpanzee.

   There were 9 trees within five steps of the most-parsimonious tree. With the addition of the 198 bp tRNA fragment to the ND4 sequence, resolution within the hominoid clade is improved, although the human-chimpanzee-gorilla trichotomy remains. Resolution in other parts of the tree remains relatively the same as for the 458 bp fragment.

4. **ND4 + tRNAs + ND5: 898 bp Fragment.** A parsimony analysis of the entire data set yielded two equally most-parsimonious trees (1,157 steps), one uniting chimpanzee and human, the other uniting chimpanzee and gorilla; all other relationships were as in the distance tree.

Within five steps of the most-parsimonious trees, five trees were found.

The consensus tree is fully resolved but a human-chimpanzee relationship was supported in only three of those trees. Interestingly, only the branch uniting the orangutan to the hominines received less support as compared to the 656 bp fragment five-step consensus tree. Overall there was a marked improvement in resolution with the larger data set.

Resolution is improved over the 656 bp fragment except in the placement of the orangutan with the hominines. Compared to the entire 898 bp data set, the five-step tree of ND4 + ND5 has less resolution within the hominines but the relationship between the hominoids and the macaques is more strongly supported by the ND4 + ND5 data set.

An analysis of the 198 bp fragment that includes the three tRNAs suggests that their contribution to instability is minimal. A single most-parsimonious tree was found (154 steps) that resolved relationships within the hominoids and macaques identical to those of the distance tree. This tRNA tree, however, placed the macaques outside the other taxa to yield: (macaques (tarsier (squirrel monkey + hominoids))).

There are 254 trees five steps away from the shortest tRNA tree and, not unexpectedly, the consensus tree exhibits relatively little resolution.

## Near — Most-Parsimonious Trees

The collection of trees within 1% of the length of the most-parsimonious tree has been used as a method to investigate the cladistic structure of sequence data and was also examined in this study.

The number of trees recovered within 1% of the shortest tree(s) did not vary noticeably among data sets despite the fact that larger data sets had trees of longer branch lengths. Only two clades, hominines and hominoids, consistently gained cladistic support as the size of the data set increased. All others showed decreasing support or no marked change. Some clades (anthropoids and macaques) were well defined no matter what size the data set. Interestingly, a global parsimony analysis shows strong resolution within the macaques across the four primary data sets, but no resolution is seen within the hominines.

## Summary of Global Parsimony Analysis

Numerous cladistic groupings are consistently revealed in a global parsimony analysis. With the exception of the smallest data set, the first 225 bp fragment of ND4, global parsimony analysis of each of the other data sets was able to resolve all relationships consistently except for the trichotomy within the hominines. This is also true of the combined ND4 + ND5 fragment of 700 bp, and, as noted above, even the most-parsimonious tree of the tRNAs was able to resolve most of the relationships.These results indicate that this relatively small 898 bp fragment of mtDNA contains significant phylogenetic signal for taxa whose divergences extend as far back as 50 to 60 million years, that even "noise" produced by the high transition rate of mtDNA cannot obscure the signal, and that a global parsimony analysis is sufficiently powerful to resolve a consistent cladistic pattern even in data that are generally thought to be "noisy." These conclusions are reinforced when the collections of near-mostparsimonious trees are examined). In general, there is improved resolution as more data are added, but the trend is not always strong. Most importantly, a consistently strong cladistic signal is found in all but the very smallest data sets. When the structure of trees within 1% of the most-parsimonious solution is examined, however, far fewer of the relationships are identified by a strong cladistic signal. Only the monophyly of the macaques and of the anthropoids is consistently revealed across all four primary data sets. As noted earlier, furthermore, saving 1% of the most-parsimonious trees does not produce a clear convergence in the phylogenetic signal as the data set is enlarged. This results from the fact that in each subset of the data there are many trees within 1% of the minimum-length tree, including some of extremely variant tree topology.

## Transversion Parsimony Analysis

Does the large bias for transition substitutions within mtDNA increase the "noise" and lead to instability of the phylogenetic signal within the primate data set? Can the signal be improved by examining transversions alone? And, does the stability of a phylogenetic signal inferred from a parsimony analysis of transversions improve with increasing sample size? These questions were investigated by repeating the above analysis using transversion differences.

1. **ND4: 225 bp Fragment.** An analysis of the first 225 bp of the ND4 fragment produced two equally parsimonious trees (79 steps). In both trees, the macaques were not fully resolved, with only the Japanese-rhesus macaque clade being held in common. The major difference between the two trees was found in the placement of Homo: in one tree it was with gorilla, in the other with chimpanzee.

   There were 681 trees within five steps of the most-parsimonious solutions. Their majority-rule consensus shows very strong support for three clades, the anthropoids, catarrhines, and macaques, but fails to resolve the relationships of any other group.

2. **ND4:458 bp Fragment.** A transversion parsimony analysis of the entire ND4 fragment yielded two equally parsimonious trees (196 steps). One is identical to the distance tree and the other failed to resolve a trichotomous relationship at the base of the macaques: (Barbary, crabeating, Japanese + rhesus).

   Although 114 trees were found within five steps of the most-parsimonious solution, the majority-rule consensus shows a marked improvement in resolution compared to the 225 bp fragment. Clades are resolved within the hominoids, but not within macaques.

3. **ND4 + tRNAs: 656 bp Fragment.** A transversion analysis of the ND4 fragment along with the tRNAs produced two equally parsimonious trees (239 steps) that have identical structure to the two most-parsimonious trees of the preceding analysis. There were 59 trees within five steps of the two most-parsimonious trees. Compared to the 458 bp fragment, some additional, slight resolution is apparent in the hominines and the macaques.

4. **ND4 + tRNAs + ND5: 898 bp Fragment.** A transversion parsimony analysis of the entire data set yielded a single most-parsimonious tree (381 steps). It has the same structure as the distance tree.

   For the transversions in the total data set, 46 trees were found within five steps of the most-parsimonious tree. Although there is no resolution within the hominines or macaques, compared to the previous consensus trees, there is a marked increase in support for the hominines and catarrhines.

5. **ND4 + ND5 Fragments Compared With the TRNA Fragment.** In a transversion parsimony analysis of the ND4 and ND5 fragments combined (700 bp), a single most-parsimonious tree was obtained (338 steps) that is identical to the distance tree. A transversion parsimony analysis of the three tRNAs (198 bp) found seven equally parsimonious trees (42 steps). The consensus tree for these showed the chimpanzee united with the human in 77% of the cases, but with no resolution among the other hominoids, including the gorilla. The Japanese and rhesus macaques were clustered as sister-species, but no other nodes were resolved within macaques, and the squirrel monkey was united with the macaques.

Transversion analysis of the 700 bp coding region finds 66 trees within five steps of the most-parsimonious tree. Their majority-rule consensus tree is very similar to that of the entire 898 bp data set except that the great apes are much less supported (76% versus 91%). Not unexpectedly, there are a large number of trees within five steps of the most-parsimonious solutions for the TRNA fragment. The majority-rule consensus for 3,000 trees (computer memory was exhausted; many more could have been found). Only the macaques as a group are well defined.

## Near—Most-Parsimonious Trees

It summarizes majority-rule consensus tree support for all trees with lengths within 1% of the most-parsimonious transversion solutions. A consistent pattern is observed: except for the TRNA fragment, virtually all of the "deep" cladistic events are strongly supported in each of the five primary analyses, and this also includes the shortest 225 bp fragment. In contrast, strong resolution of the "more recent" cladistic events, including those within the macaques and hominines, is lacking.

## Summary of Transversion Parsimony Analysis

The most general conclusion to be drawn from a comparison of the global parsimony analysis with the transversion parsimony analysis is that the use of transversions, by themselves, reveals a strong cladistic signal for the relatively older clades. This is especially evident when examining those trees having a length within 1% of the most-parsimonious solutions, which in most of the data sets included all those trees within two or three steps of the minimum length solution. Since many more trees were examined in the five-step analysis, support for some of these deeper branches declined, but in general remained fairly strong. None of these methods of analysis, when applied to trans version differences alone, were able to reveal a strong cladistic signal within the hominines or within the macaques.

## Global Parsimony of First and Second Codon Positions

Does the high rate of transitions in third positions contribute noise and therefore result in decreased stability of the phylogenetic signal? This question was investigated by a global parsimony analysis of the ND4 and ND5 fragments in which the third positions were excluded. The analysis yielded a single most-parsimonious tree of 519 steps. There were only two differences from the distance tree. The first had the chimpanzee united with the gorilla, but the most unexpected result (given the preceding analyses) was the closer relationship of the gibbon, rather than the orangutan, to the great apes. There were eight trees within five steps of the shortest tree. The consensus tree showed little resolution within either the hominoids or the macaques, but it did resolve the monophyly of both of these clades and clustered them relative to the squirrel monkey. This analysis suggests that attempting to enhance the phylogenetic signal by eliminating transitions in the third position was unsuccessful. It had no effect on improving the resolution of the older cladistic events, which were strongly resolved by global parsimony, and not unexpectedly, it led to a decrease in the signal within the macaques.

## Amino Acid Replacements: Substitution Parsimony

Amino acid replacements are often taken as being more informative of relationships than nucleotide substitutions, especially for more distant branching events. The structure of the phylogenetic signal of replacements was examined by parsimony analysis of the numbers of substitutions that account for amino acid differences among the taxa. A transformation matrix for vertebrate mtDNA provided by D. L. Swofford was used to convert amino acid differences to substitution differences and then each tree was evaluated. A single most-parsimonious tree of 387 steps was found for the 231 codons of the ND4 and ND5 fragments arranged tandemly. Relationships were similar to those of the distance tree except within the hominoids where gorilla and human were united and the gibbon (not orangutan) was the sister-group of the hominines. There were 49 trees within five steps of the most-parsimonious tree. The majority-rule consensus tree shows relatively little resolution, either within the macaques or the hominoids, thus suggesting that this method is less effective in revealing phylogenetic signal than is a global parsimony analysis.

## Bootstrap Analysis

1. **ND4: 225 bp Fragment.** A bootstrap analysis of the first 225 bp of ND4 produced a majority-rule consensus tree identical to the distance tree except for chimpanzee being united with gorilla. The latter relationship means little, however, given that all three alternatives within the hominines are about equally supported. Clades having substantial support in this data set include: (1) the macaques and their inclusive groupings; (2) the great apes; and (3) the anthropoids.
2. **ND4: 458 bp Fragment.** A bootstrap analysis of the entire ND4 fragment produced a tree very similar to the preceding, except that human and chimpanzee are narrowly united. Support for each of the clades has generally improved, especially for the catarrhines, the hominines, and the macaques.
3. **ND4 + tRNAs: 656 bp Fragment.** The majority-rule consensus tree for this data set is identical to the preceding. In general, support for each clade increases, although the catarrhine clade is found in only 50% of the replicates.
4. **ND4 + tRNAs + ND5: 898 bp Fragment.** The majority-rule consensus tree for all the data keeps the same pattern of relationships as the preceding, except for uniting chimpanzee

and gorilla (50% of the replicates). Bootstrap support for the other clades has increased, sometimes substantially (e.g., hominoids, catarrhines).

5. **ND4 + ND5 Fragments Compared With the tRNA Fragment.** A bootstrap majority-rule consensus tree of the protein-coding region shows a similar pattern to that of the distance tree. Levels of support are close to those of the 898 bp tree. Bootstrapping the 198 bp of the rRNAs suggests that they contribute less noise to the data than might be expected, except at the base of the anthropoids where the squirrel monkey is united with the hominoids. The tRNAs, however, strongly support the monophyly of the macaques and a rhesus + Japanese macaque relationship.

## Summary of the Global Parsimony Bootstrap

Not unexpectedly, the phylogenetic signal, as measured by the percentage of congruent bootstrap replicates, improves as the sample size increases. Thus, for the 898 bp fragment, six clades were supported more than 97% of the time. Nevertheless, if one preferred to interpret the "significance" of these results stringently by collapsing all branches that failed to meet a 90 or 95% criterion, the bootstraps of characters for each subset of the data would reveal a cladistic structure not very much different from that seen in previous analyses.

## Transversion Parsimony Bootstrap

1. **ND4: 225 bp Fragment.** A bootstrap of transversions contained in the first 225 bp of the ND4 fragment produces a majority-rule consensus tree much like the distance tree except for the lack of resolution at the base of the macaques. The monophyly of the macaques, the catarrhines, and the anthropoids is strongly supported by a bootstrap analysis of these data.
2. **ND4: 458 bp Fragment.** A bootstrap of the entire ND4 fragment produced a tree identical to the distance tree. Relative to the 225 bp fragment, the pattern of support varies. Clades within the hominoids are better supported with this expanded data

set, although the monophyly of the catarrhines is much less so. Support for a human + chimpanzee relationship has increased to 66% of the replicates.

3. **ND4 + tRNAs: 656 bp Fragment.** The relationships implied by this data set are the same as in the preceding analysis. There is an increase in the support of all clades except that of the catarrhines.
4. **ND4 + tRNAs + ND5: 898 bp Fragment.** A bootstrap of the transversions in the entire data set yielded a tree showing no changes in cladistic structure from the preceding analysis, but with a general increase in the cladistic support at most nodes.
5. **ND4 + ND5 Fragments Compared to the tRNA Fragment.** A transversion bootstrap analysis of the combined ND4 and ND5 sequences once again produced the same tree as did using all the data and, in general, with approximately the same level of support. An analysis of transversions in the tRNAs shows that this data set retains much of the phylogenetic structure present in the protein-coding regions, but virtually all of the nodes are poorly supported. In the tRNA data set, the squirrel monkey clusters with the macaques, showing ambiguity at the base of the anthropoids.

As with the global parsimony bootstrap analysis, there was a general increase in support for the nodes as sample size was increased. Neither bootstrap analysis provided significantly better support than the other. Although the transversion bootstrap produced much better resolution within the hominines, the global bootstrap analysis resolved the relationships within the macaques much more convincingly. Both did about equally well with the other parts of the tree.

## Informativeness, Noise and the Primate Data Set

A major concern when analyzing any systematic data is to assess the extent to which they are phylogenetically informative. Systematists are also interested in knowing whether a particular analytical method is effective in maximizing informativeness and thus results in phylogenetic hypotheses that are more strongly supported than those produced by other methods. These problems are especially acute with sequence data, not only because of the large numbers of characters that are normally generated, but also because the evolutionary dynamics of DNA are such that a large amount of noise (i.e., homoplasy) would not be unexpected.

How might the informativeness and noisiness of the primate data set be evaluated? One approach might examine sets of near most-parsimonious trees and judge informativeness by the degree to which the cladistic structure is corroborated across those trees. A second method might examine consensus trees produced by resampling of the data themselves and ask to what extent each node is supported; that is, in what percentage of the trees forming the consensus does a given node occur? Both methods examine the structure of data, but in different ways.

A consideration of the amount of support for each node as seen in the majority-rule consensus trees permits a detailed description of the phylogenetic signal contained in the data. In the global majority-rule consensus of trees within five steps of the most-parsimonious solutions, the relationships within the macaques are fully resolved using the largest data sets. The hominines always cluster relative to the orangutan, and these two groups cluster relative to the gibbon. The major lineages of the tree—hominoids, catarrhines, anthropoids, and macaques—are also well supported by the largest data sets. In the transversion majority-rule trees, there is little or no phylogenetic signal within the macaques or hominines, but the data unambiguously resolve all other clades, especially using the largest data sets. Taken together, these two analyses reveal a consistent phylogenetic signal for all clades except those within the hominines.

Bootstrap analysis constitutes a second approach to characterizing the amount of phylogenetic signal in the data. Bootstrap sampling of the total data set reveals a strong phylogenetic signal within the macaques, and comparison with the bootstrap analysis of transversions alone leads to the conclusion that a substantial component of that signal is in fact coming from transition differences. Within the hominines, however, it is transversions, not transitions, that provide the most consistent signal. In both bootstrap analyses, the major lineages are all well supported using the largest data sets, although the phylogenetic position of the squirrel monkey is the least well-corroborated.

How do the two approaches compare? Not unexpectedly, the results using these two different methods parallel one another to a very great extent. Whenever a clade is strongly supported in a near most-parsimonious consensus tree, it is also strongly supported in the comparable bootstrap tree. Differences arise in those clades that lack strong support. Because bootstrap trees tend to be fully resolved, they have the appearance of being more informative, yet there is generally little character support underlying that increased resolution. Still, when examination of the structure of near mostparsimonious trees often yields an unresolved polytomy, bootstrap resampling of the data seems more sensitive to finding the "correct" signal, even if that signal is not especially strong. Again, one need not interpret bootstrap results in a strict statistical sense, but simply as another method of evaluating the presence and strength of signal in the data. Thus, bootstrapping of transversions across the entire data set provides resolution of relationships within the macaques, whereas the majority-rule consensus analysis of near most-parsimonious trees does not. This result generally holds for most subsets of the data, not only for relationships within the macaques, but also within the hominines.

## Phylogenetic Signal Stabilize with More Data

A critical practical problem of phylogenetic inference using DNA sequences is deciding how much sequence must be obtained before relationships can be resolved. In attempting to answer this question, factors other than sequence length per se cannot be ignored. It is generally assumed, for example, that certain segments of DNA will be more informative than others and therefore less sequence will be required to reveal a phylogenetic signal. This supposition usually entails considerations of evolutionary rate, which is discussed in the next section. Here we ask whether phylogenetic signal within the primate data set is related to data set size and which method of analysis best reveals that signal.

The global parsimony analysis of near most-parsimonious trees shows an overall increase in the strength of signal as sample size is increased. There is a marked improvement in the 458 bp tree as compared to the 225 bp tree. With respect to the 656 bp consensus tree, some nodes are slightly better supported compared to the 458 bp tree, others are not. Generally, the 898 bp and 700 bp trees have stronger corroboration than the 656 bp tree.

The transversion parsimony analysis of near most-parsimonious trees also exhibits increasing strength of signal with increasing sample size. That increase in signal, however, is confined to the deeper branches and to the hominoids. Resolution within the macaques is not affected by increases in sample size using transversions alone.

As noted earlier there is a general increase in phylogenetic signal, as determined by bootstrap support, when sample size increases. This is true for both the global parsimony and transversion parsimony analyses. Not all nodes show increased support between successive subsets of data, and not all nodes are well supported, but there is a definable convergence toward a single tree. These results suggest that bootstrap analysis may be somewhat more informative as a descriptor of the effects of sample size on phylogenetic signal than is examination of near – most parsimonious trees.

## Evolutionary Rate and Phylogenetic Signal

The degree to which a segment of DNA is considered likely to be informative is generally thought to be related to its rate of base substitution relative to the divergence times that the investigator is trying to resolve. Segments with high rates of change are assumed to be less informative for deeper branches because the accumulation of homoplasy (parallel and back mutations) erases any signal. As substitutions "saturate," genetic distances among the taxa approach an asymptote relative to divergence time and this observation has led to the conclusion that mtDNA sequence variation will be less useful in resolving relationships among relatively more distantly related taxa.

To the extent that data are available, the genes examined in this study (ND4 and ND5) might be categorized as "relatively fast" compared to many other protein-coding genes, tRNAs, or ribosomal RNA genes. Hayasaka estimated that the relationship between divergence time and the number of substitutions among taxa [corrected for multiple mutations] is linear only up to about 30 million years and that older divergence times (e.g., those deeper than the base of the hominoids) cannot be resolved "conclusively." Yet, such a conclusion is method-dependent. Even though ND4 and ND5 are "fast," the present analysis suggests their rate may have only marginal influence on the relationship between phylogenetic signal and divergence time (at least within primates). Depending upon how parsimony is applied to the data, these genes are just as capable of resolving the more ancient divergences within the primates as they are the most recent. Thus, by examining the relative influence of transitions and transversions, a reasonably consistent phylogenetic signal is revealed across the tree (relationships within the hominines constitute a slight exception; see below). These conclusions deserve to be tempered, however, by noting that transition/transversion differences do not become equal until the very most distant relationships between the prosimians and the other taxa are considered. It could be argued, therefore, that relationships within the anthropoids are not affected by a saturation effect (although saturation is evident at the base of the anthropoids).

## Resolving Power of Most-Parsimonious Trees

It is often thought within molecular systematics that parsimony methods will often fail to be informative because of the high degree of homoplasy in sequence data. Therefore, it is useful to examine that supposition by asking whether the most-parsimonious trees, based on a consideration of all of the data, were in fact accurate estimates of the phylogenetic relationships of the primates, or whether the "noise" in these "fast" evolving genes made that resolution difficult or impossible.

The answer to this question may be surprising to some: the relationships of the primates were consistently revealed by the most-parsimonious trees of each of the data sets. In general, the only ambiguity involved resolution of relationships within the hominines, which has been a long-standing problem that even very large amounts of data have had difficulty in resolving. The most parsimonious solution of the smallest data set, the 225 bp fragment, also showed a trichotomy at the base of the anthropoids, yet was still remarkably effective in resolving relationships among the other taxa.

It therefore seems that a straightforward parsimony analysis of unweighted data, including both transitions and transversions, is an effective method for estimating what appears to be the "correct" tree, even in subsets of the data. It may be that because most of the data are not saturated with transition substitutions, there remains relatively little noise in the data and, therefore, parsimony is effective in revealing the signal. On the other hand, it may be that a simple parsimony approach is in fact a very powerful method and can discern phylogenetic signal, even in data having a strong component of noise.

## Parsimony and Phylogenetic Inference Using DNA Sequences

Many workers have long advocated parsimony analysis as a method of phylogenetic inference for DNA sequence data. At the same time, there has been concern within the molecular systematic community that once the algorithmic problems have been solved – and large advances have been made in recent years – there will still remain important questions as to exactly how parsimony procedures are to be applied to sequence data. The focus of most applications of parsimony has been to find the most-parsimonious tree or trees. Yet, systematic (comparative) and functional analyses of DNA and its evolution have established a body of knowledge that creates a broad spectrum of choices and challenges for the systematist. Included in this body of knowledge are the observations that the frequency of transitions and transversions are not always equal; third position changes are more frequent than first or second; and secondary structure configurations can bias the frequency and direction of nucleotide change.

The investigator is thus faced with procedural difficulties from the outset, difficulties that raise important philosophical and empirical questions. At their most basic level, these questions involve the problem of what constitutes evidence in phylogenetic inference, a subject that has received too little attention in the literature. Do we base our best estimate of phylogenetic relationships on all the sequence evidence that is available, or, if on the basis of other evidence we have reason to think that some of the data are likely to be misleading, should those be eliminated from the analysis? If the latter, what are the criteria for choosing that portion to be eliminated? Moreover, what justifications do we advance to establish the validity of those criteria? Do we treat all characters (in this case, nucleotide positions) equally ("weighted uniformly") or should some be weighted more than others (eliminating characters might also be viewed as a form of weighting)? Do we treat all character-state transformations at any given nucleotide position equally, or are they to be weighted differently? In both cases, how is weighting itself to be justified and how might weights be determined objectively? These questions can only be a prelude to a full discussion of how parsimony is to be applied inasmuch as many other issues (e.g., identifying signal, assessing topological reliability, the notion of congruence) also have a bearing on our choices regarding systematic procedure.

One theme of this chapter has been to ask what might be the best approach for applying parsimony procedures to sequence data. Because the problem involves so many complex issues, not the least of which are the philosophical tendencies of the investigator, it is unlikely that the systematic community will agree on a unified answer. If one sees the goal of systematic research to be the discovery of the one true history of life, and accepts the precept that we will never have certain knowledge of that history, then our only recourse would seem to be to accept the hypothesis of relationships which is most consistently revealed by the data regardless of which analytical techniques might have been employed (this does not imply, of course, that all methods of analysis are necessarily equally useful or appropriate). This is essentially an appeal for acceptance of stability of cladistic signal and to identify as noisy those cladistic patterns that are ambiguous or less consistently revealed.

There has been substantial support in the literature for weighting among nucleotide positions (e.g., within codons) or among alternative characterstate transformations within positions, and many weighting procedures have been suggested. Although there is a rationale for thinking that weighting will improve the resolution of cladistic signal, what is less clear is which method of weighting is "best" in any case. In principle, at least, a tree derived from weighted data should be a better or more reliable estimate of the true phylogeny than a tree derived from unweighted data, simply because the effect of weighting should be to emphasize the best available evidence. Yet, a final conclusion regarding the reliability of the resulting tree can only be reached by comparison to results based on more data and an assessment of congruence of their cladistic signal. It is possible, after all, that various assumptions about the evolutionary process that may underlie a particular method of weighting are themselves incorrect and thus may have led to spurious results. An unfortunate tendency within molecular systematics has been the application of a weighting scheme and the uncritical acceptance of the resulting tree. Currently available parsimony algorithms are designed to produce a single point solution; that is, a single most-parsimonious tree or a set of equally most parsimonious trees. Yet, it is difficult to judge from a single most-parsimonious tree whether a cladistic grouping represents a strong (stable) signal found in the data or whether the data are noisy with respect to that group's resolution (e.g., within the hominines). Thus, even if the data are weighted, questions about the reliability of the result still exist in addition to those raised by any justification of the weighting methods themselves.

The analyses undertaken in this study explore ways of revealing phylogenetic signal in sequence data through parsimony procedures. Taken together, the results of this analysis suggest that *if there is any phylogenetic signal inherent in the data, parsimony methods will find it*. The ultimate arbiter of this conclusion must be congruence with phylogenetic relationships inferred from other data sets, which in the case of the primates examined here support the conclusions of this chapter. Taken separately, each approach to parsimony reveals a significant portion of that signal, but resolution in some parts of the tree may remain ambiguous. Most-parsimonious trees based on the entire 898 bp are essentially congruent with the distance tree, there being two equally most-parsimonious trees differing only in the placement of chimpanzee with human or with gorilla; the single most-parsimonious tree using transversions alone resolved that ambiguity in favor of chimpanzee + human. An assessment of the strength of the phylogenetic signal is best described, however, by comparison of the majority-rule consensus trees derived from near – most-parsimonious solutions using global and transversion parsimony, along with those derived from bootstrap analysis of global and transversion parsimony solutions. Indeed, *a comparison using these four methods of analysis demonstrates a strong phylogenetic signal in even the smallest subsets of the data, including the first 225 bp fragment and the 198 bp tRNA fragment*. Caution must be exercised in generalizing these results inasmuch as this is only one data set, but parsimony analyses such as these appear to constitute a powerful methodological tool for resolving phylogenetic signal even in a small amount of data. Hayasaka concluded that the high rate of homoplasy in mtDNA makes the more distant relationships of the tree less reliable than those that are more recent, yet parsimony analysis revealed a strong phylogenetic signal across the entire tree.

## Phylogenetic Inference Using Sequence Data: Some Comments on Alternative Methods

There has been considerable discussion in recent years about which method of phylogenetic inference is the most appropriate for DNA sequence data. Efforts have been made to adjudicate this issue by simulation experiments in which a data set is generated based on a "known true" phylogeny and given assumptions as to how sequence evolution proceeds. Different methods are then applied to the data set to see if they recover the "true phylogeny." These simulations are perhaps useful for identifying conditions under which a given method may have limitations, but they are encumbered with philosophical and empirical problems. Because certain knowledge of phylogenetic relationships is beyond us, it is questionable whether some artificially generated "truth" can serve as the arbiter of the best method for recovering that which we cannot hope to know with certainty. The evolutionary models underlying these simulations also suffer from a critical methodological circularity: it is reasonable to assume that a detailed and empirical understanding of the mechanisms of molecular change will not be obtained unless it is based on prior hypotheses about the phylogenetic relationships of organisms. Consequently, to incorporate assumptions about the evolutionary process into a simulation procedure that will be used to choose a method of phylogenetic inference may bias our recovery of those relationships, which themselves will bias and confound our understanding of processes of molecular evolution. Perhaps as important, these simulation experiments generally take as their model of evolutionary change one that most closely applies to noncoding, randomly evolving DNA. In practice, however, the majority of molecular systematists are attempting to infer relationships from sequences that will seldom, if ever, evolve under the constraints specified by those models.

In real-world cases, we might expect that when there is a strong phylogenetic signal inherent in the data, virtually any method will recover it. When data are extremely noisy, say as a result of too little data being used to resolve very close inter nodal distances or of too much homoplasy, then perhaps no method would be expected to resolve a reliable signal. Somewhere in between these extremes we might expect that different methods will sometimes yield different phylogenetic hypotheses for a given set of data. Choice among those hypotheses should not, however, be based on an appeal to method, but on comparison and congruence with phylogenetic inferences derived from other data.

At this time, three primary methods of phylogenetic inference are applied to sequence data; namely, parsimony analysis, distance analysis, and maximum likelihood. Distance analysis has had both its defenders and critics. Tree-building algorithms that work on distance matrices are necessary for certain kinds of data, such as DNA-DNA hybridization or immunological distances, but it is questionable whether they should be applied to discrete character data such as sequences. At issue is not whether distance analysis of sequences is able to reveal phylogenetic structure, for cleary it can. More to the point is the inevitable loss of information in converting sequence data to distances, their sensitivity to saturation effects so that they must be corrected, the loss of flexibility to examine the deep structure of the data underlying the results, and the inability of most distance algorithms to examine the structure of the large suite of trees having near minimum-length fit. Taking the distance analysis of the primate data set as an example, Hayasaka did not cluster using original distances but corrected them for multiple substitutions and then clustered on those new distances. A single tree was produced by the neighbor-joining method but without an assessment of whether it was the best-fit tree. That tree, moreover, was not compared quantitatively or qualitatively to other near-minimum-length trees. Many different models have been proposed to "correct" for multiple substitutions. Yet, even though substitution probabilities can be estimated from observed nucleotide frequencies, there is no guarantee that those probabilities actually reflect the dynamics of processes that produced the observed nucleotide variation. That being the case, it remains to be seen to what extent these different models that are being applied to sequence data will influence inferred phylogenetic structure, not only of the minimum-length tree, but of those trees close to that solution.

Maximum-likelihood methods are more developed in theory than in practice and we will not discuss them extensively here. A choice among trees using maximum-likelihood methods depends upon accepting an underlying model of evolutionary change. Often, that model incorporates equal probabilities of substitutional change from one nucleotide to another, but some have been developed to account for differences in rate between transitions and transversions. Calculation of likelihood functions for trees is computationally difficult, and the method cannot be extended easily beyond cases for five or six taxa. It seems that if more realistic, and thus more complicated, models of evolutionary change are to be incorporated, computational difficulty is sure to increase.

Cedergren note that parsimony methods have virtue in phylogenetic inference using sequence data because they result "in the most economical reconstruction of mutational history, with no assumptions and with the minimum of coincidence and unobserved changes." Although it is arguable whether parsimony is assumption free (all scientific methods designed to discover pattern involve some assumptions), the strengths of parsimony analysis over other methods of analyzing sequence data are easy to enumerate. With parsimony, it is possible to use a minimum of assumptions about the processes governing molecular evolution and still obtain interpretable results. Weighting of characters or character-states involves assumptions, but that too can be accomplished conservatively such that narrowly constrained assumptions about molecular processes are avoided. Today's parsimony algorithms are fast and efficient at searching among all possible trees, thus guaranteeing that not only will the most-parsimonious tree be found, but that the cladistic structure of near – most-parsimonious trees can be examined. Moreover, with parsimony it is possible to examine and evaluate the data supporting each cladistic component on a tree. For these reasons alone, parsimony is the preferred method for analyzing DNA sequences.

# 3

# Trouble with DNA

## Introduction

The process of negotiating a plea of guilty is referred to as plea bargaining, it is in reality sentence bargaining, a negotiation over the sentence that a defendant will receive if he admits his guilt, forgoes a trial, and agrees to accept some form of punishment. It stands to reason that a defendant who pleads guilty before a trial will receive a sentence significantly less than what would be imposed after conviction at trial. After all, why else would he agree to spare the state the obligation of proving him guilty beyond a reasonable doubt? The prosecutor must weigh the rewards and risks of going to trial and decide whether justice would be served by a bargained-down admission of guilt.

As the Bronx prosecutor, Risa Sugarman, contemplated offering Joseph Castro, a handyman in the building of two murder victims, a deal in exchange for his guilty plea, she thought most of all about the two innocents viciously murdered, who were entitled to a full measure of justice. But she also had to consider the trouble with the DNA evidence, which pretrial hearings had indicated might be considerable.

Initially, Sugarman, the chief of homicide in the Bronx District Attorney's Office, believed that the DNA would clinch the case. But it later appeared that, despite the prosecutor's firm conviction that this was a very guilty man, Castro could conceivably win an acquittal at trial.

Every prosecutor's worst nightmare is an acquittal of a guilty man who has committed a heinous crime, a prospect that loomed particularly large in this case. The venue was the Bronx, and Bronx juries are notoriously friendly to defendants in criminal cases, much more so than those in Manhattan, where I worked as a prosecutor.

Besides, the likely defense here was obvious. There is an old axiom among experienced homicide detectives that the first step when a female body is found is to determine if the body had a husband. In this case, she did. David Rivera had returned home from work and found his wife and daughter murdered.

So the likely defense of Castro, the building handyman, was obvious: Who had better means, motive, and opportunity than the husband? After all, murders of wives by their husbands are common, and what marriage can stand close scrutiny without yielding up an old grudge or two capable of generating murderous rage? Even more surprising than the number of wives murdered by husbands is the number of children murdered by their parents, some 600 in the United States in a typical year. While we recoil from the prospect that a parent would murder his or her own offspring, far too many react to the stress of parenthood by lashing out at their children in uncontrolled rage. In such cases, confessions are hard to come by, because the parent knows that admitting guilt will engender more than the opprobrium of law enforcement figures. Friends, family, even fellow inmates will turn on the murderer with a passion.

Here, there was fertile ground for the defense. True, the tactic of blaming the victim, or the victim's loved one, who if innocent of the murder is certainly a victim himself, can be a risky tactic for a defense lawyer. There are few things more certain to incense a jury than a claim that a bereaved spouse is a murderer, and the potential that the tactic will backfire is great. It is a high-stakes gamble, and there are few lawyers sufficiently artful to walk that fine line between seeming to make an ugly charge against a grieving spouse and causing the jurors to start to suspect that the surviving spouse might indeed be the murderer.

In this case, one piece of evidence out of the couple's past seemed to demand a defense claim that the deaths of the mother and child could well have come at the hands of a volatile, violence- prone husband and father. By his own admission, Rivera had, a couple of years before, hit his wife and broken her jaw. This gave the defense lawyer that magic material that defense lawyers are always looking for something more to work with than mere speculation. The husband's proven potential for violence would make him fair game and perhaps provide Joseph Castro, the handyman, a way to beat the murder rap. After all, the defense would not have to prove who killed the victim, only that there was reasonable doubt that the defendant had committed the crime.

At a trial the jury would have to decide. Was the husband's story, and his accusation of Joseph Castro, entirely true? Was the defense lawyer's argument that the husband murdered his family the foulest of lies? Was the husband a master actor responsible for the deaths of his wife and daughter? Or was there reasonable doubt about who had done what?

On the day that his common-law wife, Vilma Ponce, and his two-year-old daughter, Natasha, were murdered, David Rivera was twenty years old. Rivera and his family lived near Montefiore Hospital in the Bronx, one of New York City's largest and most prominent medical centers.

The neighborhood had been solidly Jewish, but most of the Jews had long since left, and it now had more Hispanic people than any other group. Still, the fundamental character of the area remained the same. It was a good neighborhood, a place where working people tried to give their children more than they had. Rivera and his wife lived in a nice, clean apartment in a well-kept building. The apartment was full of love and care; two-year-old Natasha's many dolls were lined up neatly on three rows of shelving.

Rivera worked as a caretaker at a Bronx housing project. Ponce's life was her family. She was twenty-two years old and six months' pregnant, looking forward to the birth of her and Rivera's second child.

Once at his apartment building, Rivera took the elevator to the fifth floor and opened both locks, pushing his apartment door open. It was then he realized that something might be terribly wrong. Brown-haired Natasha was always at the front door to greet him when he came home, but that day there was only an eerie silence. Stranger yet, the chain was on the door.

The chain prevented Rivera from entering the apartment. He called out to his wife and daughter through the crack in the door. He called out each of their names more than five times. But there was no reply, or sound of movement.

Rivera went back downstairs. There, he pressed on the building intercom, ringing his apartment, hoping for a response. Then he went outside the building, looked to his apartment window, and whistled up toward the apartment.

On previous days when he had forgotten his keys, Rivera would whistle up from the street, and that would attract his wife's attention, and she would let him in the building. This day, Rivera had his keys, but the chain on the apartment door signaled that someone must be in the apartment. Why was there no answer to his whistle?

It was now approximately 5:15 P.M. Using the quarter with which he had just called home, Rivera called his mother. He thought that Vilma might have called her and that his mother might know what was going on. His mother told him she had not talked to his wife since 2:00 P.M., when Vilma had said that she did not know what they were having for dinner and the two women had talked about the television soaps. An ordinary conversation for an ordinary afternoon between a housewife and her mother-in-law.

Rivera told his mother that he was scared, that he thought something was wrong, and asked her to call the police. As soon as they hung up, his mother called the local police precinct and gave them her son's address. The officer who answered said the precinct would call 911 for her.

Rivera went back to his building, and once again he whistled up to his apartment. As he stood there, a man came out of the building.

Rivera and the man were face to face, less than ten feet away from each other. Rivera later remembered him as Hispanic, about thirty years old, 190 pounds, with medium-length hair and a mustache.

And he was covered with blood. He had blood on his wrist. He had blood on his face, covering his chin and his cheekbones. He had blood on his sneakers. The man also had an odd look in his eyes, which Rivera described as "chinky." There was nothing Asian about him; this was, according to Rivera, his way of saying that the man looked high.

As the man passed by, he looked straight at Rivera and smiled. Then he crossed the street and disappeared through a gate.

Seeing this frightful, almost Shakespearean apparition, Rivera ran back into his building. He took the elevator upstairs. No keys were needed this time to enter the apartment. And someone had taken the chain off the door from the inside.

Rivera pushed the door open slowly. He saw blood all over the apartment and ran headlong downstairs, hurtling toward the gate through which the blood-covered man had disappeared moments before. Just then the police arrived in a patrol car, summoned by the telephone call from Rivera's mother. Rivera ran toward them, and told the officers that he thought his family had been murdered.

The police went upstairs with Rivera. They entered the apartment, insisting that Rivera stay outside. Inside, there was a horrifying scene. Vilma Ponce, five-feet-four-inches tall, 125 pounds, lay on the living room floor. She was naked except for her sneakers, socks, and a black shirt pushed up high on her chest. She had been stabbed fifty-eight times. Most of the stab wounds were to her chest. Although she was half naked, she had not been raped. But she was dead.

Her little girl, two years old, all of three feet tall and thirty- five pounds, lay on the cold bathroom floor. She was fully clothed, dressed like the little girl she was, in a T-shirt, shorts, and yellow thong like shoes. She had been stabbed sixteen times, most often in the chest. She, too, was dead.

The officers went outside to talk to Rivera. They noted to themselves that he did not have any blood on him. The police told Rivera that his wife and daughter were dead, both stabbed to death. Rivera exploded in grief and anguish. In the words of one police officer, he went crazy. Most police officers place great store in their ability to judge the sincerity of a person's response to horrifying news that should be coming as a surprise. To these officers, Rivera's anguish appeared genuine.

Rivera left the building, to stay at his mother's home for the time being. The police secured the apartment and looked for fingerprints, but none other than the family's were found.

Two days later, Rivera went back to the apartment. He went back for one reason only; he had no clothes to wear other than the caretaker's uniform he had been wearing on the day of the murders. His clothes were in his apartment, and he needed them. His aunt and uncle drove him to the building.

Since the double murder, two detectives had been assigned to investigate the case. They met Rivera and accompanied him into the apartment while he gathered his clothes. Rivera left the building with his clothes; the detectives then pulled away in their car. Rivera's uncle started the car, and they were about to leave the scene when Rivera saw the man again, the same man he had seen on the afternoon of the tragedy. Today, of course, there was no blood on him. He was coming out of the same gate he had walked through on the day of the murders.

Every hair on Rivera's body stood on end. Rivera watched the man cross the street and then enter Rivera's own building. He looked different; he had shaved his head since their previous meeting. Still, he looked the same.

Rivera quickly left his uncle's car and ran after the man. Rivera approached him in the lobby and asked him his name. Joseph, said the man. Rivera then made a strange and simple request: "Can you do me a favor and smile for me?"

The man smiled, and it was then that Rivera knew for certain that this was the man who two days earlier had been covered in the blood of Rivera's own family. He left the building and had his uncle drive him immediately to the local police precinct.

At the precinct, the police showed Rivera photographs of various people who had previously been arrested. Rivera identified a man named Joseph Castro from those pictures, and told the police that Castro was the man Rivera saw on the day of the murders, and the man who had smiled for him again, on request, that very day.

So the police investigated Rivera's story. They looked into Joseph Castro, to find out as much about him as they could. They discovered that Joseph Castro, age thirty-five, five-foot- eight-inches tall, and 160 pounds, helped out from time to time in local buildings, including Rivera's, doing odd jobs. The police found out that Castro lived across the street from Rivera. Castro's mother lived in the basement apartment; he lived next door, in the meter room.

The detectives also found at least the beginnings of a motive. Specifically, they learned that exactly one week before she was murdered, Vilma Ponce had been out on the street with her friend Barbara Troy and had pointed Castro out to Troy. That man, said Ponce, is always getting fresh with me, even though I'm pregnant.

This was more than a matter of simple annoyance. There was something about Castro that frightened Ponce, said her friend, and she would never have opened the door to him. Troy urged Ponce to tell her husband about this local man who was making unwelcome advances, but Ponce was afraid to tell Rivera, possibly fearful of provoking a violent confrontation between the two. This was, at the least, evidence of a significant link between one murder victim and Castro.

The detectives also discovered that Castro may have created his own opportunity to commit this crime. One troubling detail was the fact that he could not have expected his victim to open the door to him. But the police learned that the latch on the bottom lock to Ponce and Rivera's apartment had been put in facing in the wrong direction, rendering it useless. More interesting was that the new latch had been installed only two weeks before and that the person who installed it was Joseph Castro. Castro had been assisted by the building superintendent's nephew, Alex. At a trial, Alex could testify to Castro's role in installing the defective lock.

Beyond Castro's excessive interest in Vilma Ponce, and the opportunity created by the installation of a defective lock, there was also the issue of timing. Certainly, this had been the most opportune time to surprise the young woman in her apartment. Her husband was at work, and she had just come home. Since she had just walked in the door, Ponce presumably had not yet had a chance to throw the second lock on her door, the lock that worked, the lock not installed by Castro.

The police put together this theory about timing from the fact that on the living room couch they found a brown bag containing six dollars' worth of chicken and chuck steak, obviously dinner for the family that night. The meat and chicken were still wrapped and unrefrigerated, telling proof that Ponce had not yet had a chance to put them in the refrigerator when she was attacked. There was also Ponce's 2:00 P.M. conversation with her mother-in-law, telling her that she did not yet know what they were having for dinner that night.

So, the scenario that the prosecution would describe at trial was apparent. Castro followed Ponce home as she returned with the food that would become her family's dinner and simply pushed open the door with the defective lock. He tried to force Ponce to have sex with him; that was why Ponce was naked but for her top, sneakers, and socks, why her shirt was pushed up on her chest. Ponce resisted, and Castro stabbed her to death. Castro had also stabbed and killed Ponce's two-year-old daughter. He may have stabbed Natasha to stop her screams, or because she tried to protect her mother. No one will ever know, but the bottom line was the same Castro had murdered a two-year-old girl and her mother.

After a month long investigation, the prosecutor authorized Castro's arrest. The case looked like a relatively good one. It would be hard for the defense to explain away Castro's installation of the defective lock. Still, Rivera's admission that he had once broken his wife's jaw could throw a wrench into the prosecution's well-constructed case. Rivera's past violent act against his wife meant that the prosecution would likely have to defuse the claim that Rivera killed his wife and concocted a story about Castro to throw police off his own guilt. The dramatic story of a man's leaving the scene covered in blood, the coincidence of a second meeting the very next time Rivera was at the building, and the identification of Castro as that man, based entirely on Rivera's testimony and nothing else, were almost too theatrical to be believed. The defense could even go so far as to suggest that Rivera manipulated the lock after Castro installed it, planning all along to blame his murder of his own wife on the most likely sap, a handyman known in the neighborhood to be making unwelcome advances to Rivera's wife.

To build their case, the police and the District Attorney's Office looked for additional evidence beyond Rivera's testimony. It was during this search for evidence that the police hoped would clinch the case that DNA came into the picture.

During the investigation, in the month between the crime and Castro's arrest, Castro was interviewed by a detective from the Bronx's Fifty-second precinct. The detective, while talking to him, noticed a stain on the band of Castro's watch that looked like dried blood. He took Castro's watch and watchband as evidence.

The blood on the watchband was subjected to DNA analysis to find out whether it had come from one of the two murder victims, from Castro, or from some other source. It was tested by the Life codes laboratory in Westchester, New York, the first private laboratory in the United States to do RFLP testing for criminal identification. The initial results suggested that this was the evidence that would win the case for the prosecution.

The results were compelling. Lifecodes concluded that the DNA pattern of the blood on the watchband matched Vilma Ponce's pattern at three different locations on the DNA ladder, at three different chromosomes. The laboratory stated that the frequency of this matching pattern was one in 100 million people in the Hispanic population. The focus of the case then shifted to a whole new terrain.

The Bronx criminal courthouse sits in the shadow of Yankee Stadium, and it sees more murder cases each year than there are home games at Yankee Stadium. Until now, this case had taken place on sad, awful, too familiar ground. The Bronx county courthouse is certainly a vale of tears, and it hears tales of homicide every day, the sad stories of the people's lives suddenly cut short for no rational reason.

At most homicide trials, be they in the Bronx or elsewhere, the evidence follows a familiar pattern. There is the testimony of eyewitnesses, police officers, a detective, the deceased's next of kin, and the doctor who performed the autopsy. The eyewitnesses, sometimes working people, sometimes drug dealers or prostitutes, tell the jury what they saw and identify the defendant as the person who committed the crime. The police officers testify to the defendant's arrest and to any evidence the defendant had on his person, such as a gun or a knife, that links him to the crime. The detective testifies to the defendant's confession, if there was one. The next of kin testifies to the identification of the body at the morgue, and provides some detail to humanize the deceased, so the jury will think of the fate of the victim as well as what awaits the defendant if they convict him. The doctor testifies to the cause of death.

This testimony is the simple, usual terrain of homicide. It is straightforward and often prosaic, yet at the same time highly dramatic, for it is by definition a story of life and death. It reminds everyone who hears it of the uncertain and unpredictable nature of our existence.

The Bronx criminal courts are used to bruising battles between prosecutors and defense attorneys. In this case, though, the figurative terrain of the dispute would be very different. The familiar field of controversy was replaced by an entirely new battleground, with totally unfamiliar forces wheeled up to fire on each other. It would still be a bruising battle, but now it would not be contradictory eyewitnesses challenging each other's testimony but some of the country's leading scientists. It would be a fight that would have reverberations well beyond the courtroom where it took place and briefly shake the newfound faith of many in the criminal justice system regarding the technology of DNA analysis.

All this took place even before the trial began. The judge had ordered that witnesses testify before him to help him decide whether this new DNA evidence had gained sufficient acceptance among scientists to be included in the evidence a jury would be allowed to hear at the trial.

At first, everything appeared to be just right with the DNA. The prosecution's first witness, Dr. Richard Roberts, worked closely at the time with Nobel Prize winner James Watson. Watson had become one of the world's best-known scientists when his unprecedented book, The Double Helix, provided the inside story of his discovery with Francis Crick and Maurice Wilkins of the helical structure of DNA in 1953. Watson was the director of the Cold Spring Harbor Laboratory, one of the world's most prominent biological laboratories, and Roberts was assistant director of the laboratory. This was a witness promethean by association. In years to come, Roberts too would win the Nobel Prize.

This impressive witness lent his full support to the new DNA technology. It was not possible, said Roberts, for DNA analysis to result in a false positive. It was hard to imagine a stronger general endorsement of DNA analysis than that provided by Roberts in court, an affirmative statement to the effect that this evidence could never contribute to a wrongful conviction.

Beyond endorsing DNA generally, Roberts also seemed to embrace the results in this case, to support the reliability of the results that implicated Joseph Castro. He testified that he had reviewed the X-ray films that embody the DNA results. Initially, he said, he was troubled by one of those films, which was not perfect. But a second and third set of experiments were carried out, and these gave a perfectly clean interpretation. Roberts was not specifically asked and did not specifically state that there was a DNA match here. Still, his testimony seemed to affirm the scientific reliability of the specific results generated by the DNA laboratory in this case.

Little did the prosecution realize that this would be the high point for their DNA evidence. From this point forward, there was nothing but trouble ahead. The groundwork for that trouble had been laid the month before, on February 16, 1989, at a conference on the use of DNA in the criminal field held at Roberts's own laboratory at Cold Spring Harbor. The conference was sponsored by the Banbury Center at Cold Spring Harbor, which hosts multidisciplinary conferences and workshops on social issues relating to molecular biology. A scientist from Life codes, Dr. Michael Baird, spoke and showed films of casework done by his laboratory.

Attending the presentation was Dr. Eric Lander, then thirty two. Lander was an MIT scientist, a former Rhodes scholar, and winner of a MacArthur "genius" fellowship, an award of $205,000 given without application or any strings to a person whose work deserves support. Lander was disturbed by aspects of the Life codes presentation. Most particularly, he was troubled by the possible inadequacy of the laboratory's controls.

Lander stood up at the meeting, and confronted the Lifecodes scientist about the adequacy of the controls used by the laboratory to guarantee the reliability of the lab's results.

During a break in the conference, Lander was approached by Peter Neufeld, then thirty-nine. Neufeld was one of Castro's two court-appointed lawyers for the DNA hearing. He had agreed to handle the DNA hearing with another attorney, his old friend, Barry Scheck, then forty.

Neufeld and Scheck, who had been brought into the Castro case specifically to handle the DNA hearing, were two of a very small group of lawyers who had knowledge about DNA evidence at this time. They were attending the "by-invitation-only" Banbury

Conference as representatives of the defense bar who had previously worked on cases involving other forms of scientific evidence.

Neufeld showed Lander the evidence from the Castro case, the pieces of film that were the final result of the DNA analysis process, known as autoradiographs. Lander was unimpressed with the laboratory's results, and told Neufeld as much.

As far as Lander was concerned, that was the end of it. He would go back to his genetics laboratory, and the lawyers would return to their offices and their trials.

But Scheck and Neufeld believed that they had found their expert witness, their vehicle to undermine the DNA evidence. In Lander they saw a kindred spirit, a questioning scientist whose skepticism could translate to some measure of success for the defense side at the DNA hearing.

Certainly, Scheck, Neufeld, and Lander were almost peas in a pod. All three were New York born, well educated, highly intelligent, and extremely argumentative. All tended to speak in sweeping, categorical terms, and none lacked the arrogant certitude that he was right no matter how the rest of the world might feel. And a defense assault on DNA would certainly require that certitude, for the defense lawyers in particular were arrayed against a criminal justice establishment that repeatedly asserted that there could be no such event as a false DNA match. To challenge DNA was to stand out on a limb and challenge the tree. But that was what Scheck and Neufeld prepared to do.

Several days later, Neufeld called Lander and asked him to testify for the defense. Lander declined. He had other major commitments taking much of his time. Lander told Neufeld that he needed to become involved in a court case as much as he needed a hole in the head. Besides, Lander doubted that the problems with Life codes 's work could be that egregious.

While Lander turned down Neufeld's request that he testify as an expert, he did agree to educate Scheck and Neufeld. The defense lawyers sent Lander daily copies of transcripts from the courtroom testimony, and the scientist reviewed that testimony, suggesting questions to the defense lawyers for their cross examination of the witness from the Life codes DNA laboratory.

As Lander reviewed transcripts from the courtroom, he did not like the answers that the Lifecodes scientist was giving to the questions he had suggested. To Lander, they sounded like excuses for inadequate science. As his frustration mounted, Lander agreed to testify for the defense. He ultimately donated 350 hours of work to the case, including six days of testimony and a fifty-page expert witness report. It was Lander's testimony, his criticism of Life codes 's procedures, delivered with the full force of his personality and intellect, that above all determined the outcome of the Castro DNA proceeding.

Aided by Lander's suggested questions, the defense won important concessions on its cross-examination of the prosecution's principal DNA witness, Dr. Michael Baird, of Life codes, whom Lander had confronted at the Cold Spring Harbor conference. On cross-examination by Barry Scheck, Baird admitted that Life codes 's reports contained language implying that the laboratory used objective, mathematical standards in determining whether or not there was a DNA match. On the witness stand, he conceded that no such calculations were made and that DNA matches were declared simply based on visual observation.

To the feisty Scheck, the significance of Baird's admission was dramatic, and he threw both arms straight up in the air, pronouncing through his gesture that he had just scored a touchdown, as if the importance of the confession should have been apparent to all. "I mean, am I crazy?" he asked the judge. In the contentious atmosphere of the courtroom, the prosecutor asked Scheck if he "would like an answer to that question." The judge intervened to stop the hostilities.

The defense had scored a small but significant victory, an admission that the laboratory's printed materials were apparently misleading when they stated that the lab declares DNA matches objectively, and that in truth it made the determination subjectively.

This alone might not have been sufficient to block the admission of the DNA evidence, although it would certainly have been useful at trial in undercutting the lab's credibility with the jury. More would be needed to keep the DNA out of the trial altogether and more would be supplied.

The bombshell came from Dr. Lander. Going beyond finding ambiguities and problems in the DNA evidence, he told the judge that the DNA evidence here "would indicate" that the blood on Castro's watchband did not come from the crime, that the analysis could show Castro "absolutely innocent."

His testimony was powerful and detailed, identifying a number of faults he had discovered with Life codes 's procedures. Lander completed his testimony on April 26, 1989. Soon afterward, at the end of April, Lander and Roberts met at a professional conference at Cold Spring Harbor on various scientific issues unrelated to criminal justice. The two adversaries in the Castro case discussed the case, and Lander gave Roberts a copy of his fifty-page report. Roberts reviewed Lander's report, and decided that something had to be done. "I hadn't really seen the evidence in great detail before, and I quickly became rather concerned," Roberts later said.

Four days later, while in Boston on other business, Roberts visited Lander in nearby Cambridge. Roberts suggested that the experts on both sides meet and discuss the issues as scientists, with "none of this lawyerly talk." A meeting, without lawyers, was scheduled in New York for May 11.

The conclave led to an extraordinary two-page statement, issued jointly under the signatures of two prosecution witnesses and two defense witnesses. The scientists concluded that the DNA results in the Castro case were not sufficient to determine that the DNA did match or that it did not match. They agreed that the DNA results were inconclusive.

This was a dramatic departure from the prior testimony by the witnesses on both sides. Roberts, of course, had previously testified that the DNA results looked perfectly clean to him. And Lander had testified that the results would indicate that the blood did not come from the crime. But in their joint statement, the experts concluded that the DNA evidence neither implicated Joseph Castro nor cleared him. This statement was later endorsed by all the expert witnesses in the case except Life codes 's Baird, who stuck to his view that the evidence implicated Castro.

The experts were troubled because at one location on a DNA print there were two faint, extra bands present in Vilma Ponce's DNA that were not present in the DNA from the watchband. Life codes had discounted these bands, viewing them as the likely result of bacterial contamination rather than genuine DNA bands, and had focused on the match between the more clearly visible DNA bands. Lander, on the other hand, believed the faint bands to be human DNA bands; that had been the basis of his conclusion that the DNA on Castro's watch did not match Ponce's. The experts agreed that the laboratory should have performed further tests regarding the source of those bands before declaring a DNA match. They decided that Life codes 's failure to do so rendered the film inconclusive at that particular DNA location.

The experts also criticized Life codes 's declaration of a DNA match because they believed that the laboratory had violated its own rules for declaring a match, that it did not use objective criteria. The criticism here went beyond a failure to perform the requisite mathematical calculations. According to the joint statement, those calculations, when actually performed, showed that the DNA patterns from the blood on the watchband and from Vilma Ponce's DNA, while similar, were sufficiently different that they fell outside Life codes's own mathematical standards for the confident declaration of a match.

The joint statement also concluded that Lifecodes's approach overstated the numerical significance of a DNA match. The laboratory had used very broad standards in determining the existence of a match. It had then used very narrow principles for calculating the numerical significance of a match. According to Lander, this was like catching DNA with a ten-foot-wide butterfly net and then showing the difficulty of that feat by demonstrating how hard it is to catch DNA with a six-inch-wide butterfly net.

After the experts issued their joint statement, the defense called Dr. Roberts back to the witness stand. Roberts now told the judge that he could not say whether the DNA in this case matched or not and that most scientists would share his view that the evidence against Castro was inconclusive.

Against this background, the judge ruled that the jury could not hear Life codes 's conclusion that the blood on Joseph Castro's watchband was that of Vilma Ponce. This was the principal DNA evidence, and it would not go to the jury.

Justice Gerald Sheindlin, writing with an eye to the future, went out of his way to emphasize his finding that DNA testing is highly reliable and that his ruling rejecting the DNA evidence in this case was premised primarily on the laboratory's failure to conduct additional tests on the faint bands to determine whether they were human DNA or bacteria. Judge Sheindlin did not accept or reject the experts' conclusion that the laboratory violated its own standards for declaring a match, but he did find that the DNA numbers were exaggerated and wrote that he would have been forced to preclude or substantially reduce those numbers had he allowed evidence of the DNA match.

The judge did rule that the jury would be allowed to hear that DNA analysis showed the blood on the watch was not Castro's. Scheck and Neufeld had not challenged this DNA evidence at the hearing. This evidence could have had some value as proof against Castro at trial after all, why was he walking around with someone else's blood on his watch? but it would have had nowhere near the power of the excluded evidence, the barred conclusion that the blood on the watch was Vilma Ponce's.

Although the evidence of a DNA match had been excluded, the case could go forward, based primarily on Rivera's testimony and that of the young man who had been with Castro when he installed the lock that later proved defective. The prosecution and the defense had reached a momentous point. David Rivera would now confront Joseph Castro in court.

This confrontation was to take place immediately before the trial began, at yet another pretrial hearing. Castro was no longer represented by Scheck and Neufeld, his special DNA counsel, but by his original, court-appointed attorney. As Rivera took the witness stand, his rage at Castro was palpable. All present in the courtroom could feel it. The prosecutor asked Rivera a question about the building he lived in, and Rivera could not even bring himself to respond. There was a feeling in the air that Rivera might come out of the witness chair and physically attack Castro.

The judge told Rivera to look just at the DA, not at Castro. Rivera then recounted his story to the court. DNA or no DNA, it was evident that this was a man who had nothing to do with the murder of his wife and daughter, a man whose whole life has been wrenched by an evil act. It was also apparent that Rivera firmly believed that Castro was the man who murdered his family.

Credibility is key in the world of the courtroom. Credibility revolves about the consistency of a witness's story with those of other witnesses and about feasibility, whether or not the testimony affronts common sense. But judgments about credibility are also heavily influenced by intangibles, and Rivera was a man whom a jury would most likely believe.

The time had come for the prosecutor to decide whether to offer a plea bargain. In many respects, the case against Castro was still strong. Rivera would likely be an emotionally compelling witness at trial, but he was still vulnerable to cross-examination, and to the defense's speculation that he might have been the killer because he had broken his wife's jaw. And this was the Bronx, a tough county in which to get criminal convictions.

The prosecutor and the judge offered Castro a sentence of twenty years to life if he pled guilty to both murders without going to trial. The loss of the DNA evidence had cost the prosecution many years off Castro's sentence. Beating the DNA had saved Castro years in prison.

As he pled guilty, Castro admitted that he had stabbed each victim repeatedly and that he intended to kill both woman and child when he stabbed them. He also resolved a central question in the case. Castro confirmed that the blood on the watch came from the crime, essentially admitting that Lifecodes's conclusion was correct.

This provided some small measure of vindication for the DNA laboratory. But Castro's plea of guilty was virtually ignored in the publicity that followed the rejection of the DNA evidence in his case. To this day, much of the lingering popular perception that DNA evidence is somehow flawed has its origins in the press coverage of the Castro case.

Before this case, the press had been positively infatuated with DNA evidence. This reflected in part some of the overblown claims of the proponents of DNA, which included the claim that DNA analysis could never yield a mistake, a questionable claim about any system subject to the vagaries of human error. When the bubble of infatuation burst, the fallout was harsh. Where before there had been a vision of perfection, now there was the taint of "junk science."

This syndrome was most pronounced in the *New York Times*'s reporting on DNA testing in the aftermath of the Castro case. On

January 29, 1990, the *Times* ran a front-page article titled "Some Scientists Doubt the Value of 'Genetic Fingerprint' Evidence," apparently inspired by the fallout from the Castro case, although it never specifically mentioned that case. The article, by *Times* reporter Gina Kolata, reported that "leading molecular biologists" say that DNA analysis is "too unreliable to be used in court."

The impact on the public's perception of this new evidentiary tool was enormous, for the article suggested that a technology that in previous press reports had been universally praised as foolproof was really invalid. But this press report was more than a little peculiar, for several of the "leading molecular biologists" quoted by Kolata were known to be avid supporters of DNA analysis in criminal cases.

Sure enough, four of the seven "leading molecular biologists" reported in the *Times* article to be supporters of the view that DNA is so unreliable that it should not be used in court joined together in a scorching letter to the editor of the *Times*. They characterized Kolata's article as a "misrepresentation" of the validity of DNA identification and stated that there is "excellent agreement among scientists that the validity of the DNA identification method is widely accepted." The biologists went on to write that the public deserves to have this powerful technology protecting its safety against violent criminals, and the wrongly accused deserve to have the technology prove their innocence. The letter pointed out that the Castro decision was based on technical aspects of the particular case and not the fundamental scientific validity of DNA technology.

Astonishingly, *the New York Times* did not print the scientists' letter. The public now moved to the view that the great promise invested in DNA was unraveling. This was a matter of substantial concern, because judges and jurors are also members of the public. It looked like the incorrect perception that DNA evidence was unreliable could be a burden to overcome in every case into which it might be introduced.

Fortunately, most judges continued to recognize the validity of DNA testing. They knew that the Castro case generally endorsed the reliability of DNA testing and excluded the results implicating Castro principally because Lifecodes had failed to conduct additional tests on faint DNA bands. As a consequence, the critical article did not do as much damage as had first been feared. After Castro, courts continued to accept DNA evidence as leading scientists continued to testify that this was an extremely reliable technology. And, for legal purposes, the Castro decision itself came to stand for that view, as well as for the possibility that a laboratory can perform DNA analysis and interpretation improperly in an individual case.

At the time of the Castro decision, there was already a very substantial body of law endorsing the reliability of DNA evidence. Defense critics argued that all those prior cases should be discounted because they did not follow a hearing as searching or extensive as that which took place in the Castro case. But in most of those cases capable defense lawyers had consulted with qualified scientific witnesses and been unable to find anything significant to challenge in the DNA evidence.

To some extent, the courts recognized that the Castro matter simply represented individual failings in one particular case. "There is a great danger of making too much of the Castro case," said Dr. Roberts at the time. "From the evidence I've seen of other cases, I don't think it is typical."

After all, it was unlikely that laboratories routinely violated their own match criteria, or that there were faint extra bands present in most cases. More likely, the Castro case represented a confluence of the reality that there were some ambiguities in the DNA evidence that the lab had not taken steps to resolve, conflicting statements by the lab regarding its own criteria, the retention of two very aggressive defense lawyers, and a remarkable expert witness they were able to persuade to join that case.

But there is no question that the Castro case heightened interest in some issues surrounding the reliability of DNA evidence that went beyond the guilt or innocence of Joseph Castro. Above all it demonstrated the need for developing appropriate standards and requiring adherence to those standards. The experts for the prosecution and defense in the Bronx case both complained about Life codes 's failure to adhere to its own policy of using objective match criteria. They also agreed that Life codes in 1989 was using statistical techniques that dramatically overstated the numerical significance of a DNA match.

Some of the concerns raised by the critics of DNA testing in the first years of such testing were related to the private and commercial nature of the first DNA laboratories. The field of DNA testing in the United States was originally developed and dominated by two private commercial laboratories, Lifecodes in New York State and Cellmark in Maryland. At the beginning, only Lifecodes offered RFLP testing; Cellmark followed shortly thereafter. Both offered their analysis to police departments and to district attorneys' offices for a fee, leading inevitably to criticism that the profit motive had driven private laboratories to rush DNA testing to market before adequate standards to ensure reliability were developed and put in place, and with overstated statistical claims about the significance when a DNA match was found.

So in 1989 there was a great deal of ambivalence about DNA testing. There had been the public relations debacle of the Castro case, and the *New York Times* article suggesting to the public that DNA testing was unreliable. More significant, there was an unquestioned need for standards to guarantee the reliability of DNA testing, particularly testing done by commercial laboratories run for profit, so that the results would not be subject to suspicion that they were monetarily motivated.

Above all, everybody involved with DNA knew that there must be no more Castro cases, or the acceptance of DNA could be irretrievably damaged. Where would the leadership come from to advance the reputation of this fledgling but already powerful technology?

Agent Presley is tall and fair haired, and has an all-American, open manner. He is a graduate of Wheeling Jesuit College, and he fulfills the expectations one would have of an FBI agent smart, serious, and reliable. Our first meeting was occasioned by his visit to New York to participate in a program at the District Attorney's Office. I was to do a mock prosecutor's examination of a DNA expert for the new lawyers in the office, to show how to question an expert witness. A defense lawyer was then to cross- examine the expert.

Like all the FBI agents, Presley was new to testifying about DNA. As much as I wanted the experience of questioning a DNA expert, Presley, who would soon testify in hundreds of cases, wanted the experience of being questioned by a prosecutor and cross-examined by a defense attorney. Long before this meeting, the FBI had committed itself to getting DNA analysis established on a firm footing. Presley was part of that effort and had volunteered to play the expert in the New York exercise.

The FBI had moved slowly and cautiously into the field of DNA testing, starting its research in 1986 but not becoming involved in actual cases until later. While Life codes and Cellmark were handling cases, the FBI was still doing experiments. The Bureau was testing DNA samples and checking its own results, to make sure that it could do DNA analysis reliably. Soon, the FBI went beyond research to performing DNA analysis on actual criminal cases, accepting cases for analysis late in 1988 and reporting its first result in an actual case in March 1989.

The FBI made DNA testing available at no charge to law enforcement authorities. Agents traveled around the country to testify to the results of DNA tests performed by scientific personnel working in the FBI laboratory, putting the quality of the FBI lab and the credibility of the Bureau squarely behind DNA. Sometimes the agents testified for the prosecution and sometimes they testified for the defense.

But the FBI's efforts extended well beyond providing high-quality DNA testing and testimony in ongoing cases. At the time, eight scientists worked full-time at the Bureau on research and training. Their work was headed by Bruce Budowle, who quickly became a guru to DNA scientists all over the country.

At the time, the FBI was subject to heavy lobbying by the two private biotechnology companies active in the use of DNA for criminal identification, Life codes and Cellmark. Each used a system with different components, although each system used the RFLP method. Both companies understood that the FBI's use and endorsement of their technology could mean enormous profits. By adopting either corporation's technology, the FBI would enhance that company's prominence in the field of DNA testing. At the same time, that company would see an increase in income, from licensing fees and the sale of their products to the FBI and other crime laboratories.

There were two fundamental differences between the RFLP systems used by Lifecodes and Cellmark. First, each used a different chemical material to initially cut the DNA. When the FBI selected a third chemical material for use in its system, both Lifecodes and Cellmark switched to using that material. Second, Cellmark used the original system developed by Dr. Jeffreys, the inventor of DNA analysis. Lifecodes used a refinement on that technique, one that was less like a fingerprint but more easily interpretable. The Jeffreys "multi-locus probe" reflected many genetic areas on one piece of film; the Lifecodes "single-locus probe" reflected DNA analysis from each genetic location on a separate piece of X-ray film. The FBI adopted the Lifecodes approach for use in its RFLP system, and Cellmark soon switched to it too.

Neither company would buck the FBI. It was not that the FBI came into DNA testing as the principal leader in the field of serology, or blood analysis (as it was in other areas, like hair and fiber analysis) – that role had always fallen to Scotland Yard, and to the large urban crime labs in the United States. But the FBI laboratory was respected, its credibility unimpeached, and its endorsement crucial to the acceptance of DNA testing by courts, District Attorney's Offices, and local crime laboratories. And to place the new technology on a solid footing, Budowle sought the guidance of leading university scientists unaffiliated with the biotechnology companies such as Ray White at the Howard Hughes Medical Institute at the University of Utah, Thomas Caskey at Baylor University, and Kenneth Kidd at Yale – listened carefully, and then performed numerous experiments.

Under Budowle's leadership, the FBI explored the boundaries of DNA technology. Bloodstains were exposed to sunlight, gasoline, motor oil, and detergent. More than 2,000 items were analyzed. In every case, the result before exposure was the same as afterward. Contamination could obliterate a DNA print, but it would not alter it. There would be a correct result, or no result, but environmental contamination would not produce a false result.

Budowle also sought to develop a means of stating the significance of a DNA match that would be both understandable by lay jury members and unimpeachable by scientist witnesses. In stating its results, the FBI specifically chose statistical methods that would, if anything, understate the significance of a match, a sharp contrast from Lifecodes's policy so that defendants got a break from the FBI's statistical methods. There would be no pretense that six-inch butterfly nets had been used in stating the significance of a DNA match.

The FBI research and training efforts also included developing techniques for DNA analysis that would be easily interpretable. Before the Castro decision, the Bureau had already decided not to use the particular test that would generate the ambiguous results in that case. After developing its own testing techniques, the FBI trained scientists from all over the United States, and the world, in its techniques. As a result, the FBI techniques are those most widely used in local crime laboratories from coast to coast today.

Under Budowle, the FBI also sponsored an effort to develop standards that could be applied to all laboratories, providing objective rules for private as well as public laboratories. This effort drew on outside scientists, independent of the FBI, including Ray White, the Utah scientist who had developed RFLP analysis for genetic diagnostics.

This group of scientists first met in November 1988, and the FBI published their guidelines for DNA analysis in 1989. The guidelines were meant to set standards that would ensure the quality and reliability of DNA testing. They addressed all the concerns raised by the joint statement of the experts in the Castro case, requiring among other things that DNA matches be confirmed quantitatively, that standards be developed for determining when DNA analysis is inconclusive or unable to be read, and that scientifically valid statistics be used for stating the significance of a DNA match.

After 1989 the scientists met again and issued further reports, but the group of independent scientists working in association with the FBI were already well on their way to establishing a set of rules to govern DNA analysis, whether done in the FBI's laboratory, at Lifecodes, at Cellmark, or elsewhere.

As the decade drew to a close, much of the broad acceptance of DNA testing that the press reporting in the aftermath of the Castro decision had undermined was established once more.

DNA's return to a state of grace was confirmed when the Office of Technology Assessment of the United States Congress issued its long-awaited report on DNA testing in 1990. The report to the Congress noted that standards were still being developed and that more work needed to be done. Its most crucial finding, however, was devastating to the DNA debunkers.

Testing of DNA is reliable, concluded the report, and questions about the validity of DNA typing are "red herrings that do the courts and the public a disservice." Like the scientists who wrote the unpublished letter to the *New York Times*, the congressional Office of Technology Assessment recognized the fundamental reliability of DNA analysis and its important role in identifying the guilty and exonerating the innocent.

After one short year, DNA was back on track. DNA testing had been fundamentally reliable to begin with; by 1990, because of the exceptional efforts of the FBI, the standards and procedures necessary to protect its reliability using the RFLP method were mostly in place. The FBI effort was already well under way before the Castro decision, but the Bronx case raised the stakes, and gave much more impetus to the need for care and quality in DNA testing. It was one thing to want to do good, reliable work. It was another matter for a scientist to know that his or her work would be subject to the closest possible scrutiny by others who want to tear that work apart.

The scientific community met the challenge of setting standards to promote reliable DNA testing. Before, DNA testing had seemed simple, inevitable, infallible; now, in part because of the efforts of critics like Lander, Scheck, and Neufeld, it was apparent that DNA analysis was highly reliable but would require care in its use and interpretation. But the standards were apparently in place to govern that process, subject to the care and scrutiny of prosecutors, defense lawyers, and scientists. New approaches to protecting the integrity of DNA analysis would also be forthcoming; the influence of the Castro case would continue well into the future, for the joint statement of the prosecution and defense experts from that case urged that the National Academy of Sciences study DNA testing, and a report eventually published by that organization would figure dramatically in future cases and controversy over DNA analysis.

Still, in the wake of the trouble over DNA, the future of the new technology was far from assured. Would prosecutors and judges stand squarely behind the new technology, even when they did not like the results it generated? Defense lawyers could pick and choose between those cases in which they liked DNA results and those in which they did not, relying on favorable DNA results and attacking DNA tests that implicated their clients. Would prosecutors do the same, or fully support the reliability of DNA testing and find ways to work around DNA results and place them in the appropriate context when they did not like them but still believed a suspect to be guilty? This question was soon put to the test in one of the most notorious criminal cases of our time, the Central Park jogger case.

# 4

# DNA Matching

## Introduction

The basic concepts are always the hardest. This is particularly true in the study of evidence. Relevant evidence is evidence that alte the probability of a fact that matters, and relevant evidence generall is admissible unless it is too prejudicial. Despite this seemingly formulation, questions of relevance relating to DNA evidence even to the most elementary concept of a "match" between two DNA samples - can be confusing.

Well established methods of molecular biology permit laboratories to compare DNA from a crime scene to DNA from a defendant. If the DNA in these samples is similar, the match usually is powerful evidence that the incriminating DNA came from the defendant. To describe the incriminating effect of the resemblance, scientists may use numbers. The numbers seen most frequently in criminal cases are numbers referred to as the match frequency. These are estimates of the relative frequency of other people whose DNA also would match the crime scene DNA according to the criteria the laboratory uses for declaring a match.

In court, the most successful objection to these numbers has been that the standard computational method does not account adequately for "population structure." Arguments targeting the criteria for de a match have not fared as well. In a recent article on the "DNA War," Professor William C. Thompson sought to correct this situation. Already a veteran of a few skirmishes in the war, Pro Thompson believes that when a defendant argues that "the match criteri should have been drawn in a narrower manner that excludes him," the finding of a match "could properly be excluded under Federal Rule 403." At the very least, he believes that the court should allow expert testimony to the jury that the match criteria are too inc In contrast, I have argued that:

> *Disputes over the strictness of particular windows or the optimal match window - when there is no such thing - may confuse and perplex the jury when it considers the probative value of a match. As a result, a has discretion to exclude this testimony.*

## Matching

The most common form of DNA analysis in criminal cases utilizes four or five "single locus VNTR probes" to produce a "multilocus genot" or, more simply, a "DNA profile." Technicians use techniques of molecular biology to excise fragments of chromosomes that begin and end with certain sequences of DNA base pairs from samples found in blood, semen, or other material containing sufficient DNA. The beginning and ending sequences are chosen so that the material they bracket tends to vary in size from person to person.

The lengths of the DNA fragments are measured by seeing how far they move, relative to DNA fragments of known lengths, through a slab of gelatinous material, under the pull of an electric charge. The procedure is known as electrophoresis. In a given time, shorter fragme (with low molecular weight) migrate farther down the electrophoretic gel than longer fragments (with high molecular weight). Fragments of approximately the same size will stop at more or less the same point, coalescing to form a spot or band. By measuring the locati of the bands it is possible to determine the approximate lengths of the fragments. The prevailing method of agarose gel electrophoresi of VNTR fragments is not sensitive enough to distinguish between fragments that are extremely close in size. Measurement error creeps in.

The likely extent of this measurement error is found through reprodu studies. For example, a laboratory may compare the fragment lengths in DNA obtained from vaginal swabs with that of DNA in blood taken from the same woman. Such studies enable the laboratory to choose a "window" wide enough to be likely to result in match when two samples come from the same source. For instance, studies conducted by the South Carolina Law Enforcement Division DNA laboratory found that corresponding bands never differed by more than [+ or -] 2.82% of their average length. The FBI reports biggest difference observed in its laboratory is less than [+ or -] 2.

Once a laboratory establishes a numerical matching rule, technicians can compare the crime scene fragment lengths to fragment lengths from a suspect. Suppose that a window of width [+ or -] 2.5% i chosen. If a visual match is declared, and if all pairs of corresponding bands in the two profiles differ by no more than [+ or -] 2.5%, then t profiles are said to match.

The match window does not have to be any particular size. Big windows are likely to result in the declaration of a match when the DNA samples come from the same person. In this case, the procedure is highly sensitive - it has a large probability of declaring a match When the samples have a common source. Along with a high sensitivity, a suitable window should yield a high specificity. Specificity refers to the probability that the procedure will not declare a match when the samples do not come from the same person. A wide window enhances sensitivity; a narrow window favors specificity. When the sensitivity is high, DNA from the same sources almost always match and there are relatively few false negatives. When the specificity is DNA from different people almost always fail to match and there are few false positives. Thus, statistical reasoning gives the operati characteristics of the matching procedure, but scientific values d not dictate the choice of a unique match window.

Once a reproducibility study shows that a window has reasonable specificity, the evidentiary analysis should focus on ensuring that the testimony about the match fairly describes the probative value of the laboratory results. Will testimony about the properties of different windows aid in this task? Are there situations where the difficulty of expressing the probative value makes it better to exclude the evidence of a match? To answer these questions, it is necessary to analyze plau ways to express the probative value of the "match." At least three possibilities exist: proportions, similarity likelihood ratios, and match-binning likeiihood ratios.

The conventional procedure for presenting incriminating DNA evidence in this country entails not just the declaration of a match, but also an estimate of the relative frequency with which other people in some population would match the crime scene DNA. This matching proportion of the population, designated as some number P, plainly has some connection to the probative value of the match. I P approaches 1, then so many innocent people would match that the match proves little; if P approaches zero, then it may prove a great deal.

It is possible to estimate P from a random sample of people from a suitable reference population. For example, consider a case in which a woman is kidnapped, assaulted, and robbed at a rest stop on an interstate highway by a truck driver who is white, and in which the DNA comparison is limited to a single band. With DNA from a suitable randcm sample of Caucasians, P could be estimated on the basis of the proportions of bands that fall into predefined intervals molecular weights.

The height of each column is proportional to the fraction of bands that lie within each "bin," which are 12%, 8%, 16%, 8%, 4%, 8%, 20%, 8%, 12%, and 4%, respectively. No bands lie outside the numbered intervals.

Suppose that the crime scene DNA band is in the center of bin 0 and that the bins are much wider than the match window. Then an estimate twelve percent of the bands from the Caucasian population would "match." Under the simplified scenario of examining a randomly chosen single band, P = .12. In this hypothetical case, the prosecution might offer testimony that the samples match and that such matches would be seen in no more than twelve percent of the Caucasian population.

## Probative Value

The best developed and most plausible theory of probative value articulated by legal scholars builds on a statistical concept known as the likelihood ratio, or, more generally, the likelihood function. The idea is elegantly simple and exceedingly powerful. Evidence has probative value with respect to competing hypotheses about the process that generated the evidence. With DNA measurements there are only two hypotheses: S, that the data arose from a common source of DNA, and -S, that they did not. To the extent that the data are mo probable under S than -S, they support S over -S. To measure this degree of support, the "likelihood" of S is defined as being proportional to the conditional probability of the data given that S is true. A convenient measure of probative value is the ratio of the likelihood for the two hypotheses:

> *As long as L is greater than 1, the DNA data support the hypothesis S. When L is less than 1, they support -S. When L equals 1, they are expected to arise as frequently when one hypothesis is true as when the other is true. Hence, they provide no reason to favor one hypothesis over the other and are irrelevant to choosing between them. Thus, the more the value of L departs from unity, the greater the prob value of the data.*

Despite its mathematical form, equation (1) is ambiguous: How should the "data" be represented? At the most detailed level, DNA data consist of two sets of numbers denoting the approximate molecular weights of the fragments from the two samples. The useful information in these two profiles involves both the observed weights of each pair of corresponding fragments and the similarity in each pair of measurements. For instance, suppose that a single-locus probe produ a band that seems to be 2328 base pairs long for the crime scene sample and a corresponding band 2352 base pairs long for the defendant's DNA, and that this is the only band to consider. The measurements are each 12 base pairs away from their mean of 2340 base pairs. The probative value of this pair of measurements depends on how rare bands in this region are, as well as the measured different of 24 base pairs (1.03%). This difference, (d), together with the mean weight x, may be substituted for the expression "data" in (1) to yield the "similarity likelihood ratio":

> *For a particular mean fragment weight r, the closer d is to 0, the bet the corresponding bands match, and the greater the value of L. For a given separation d, the rarer the measurement weights are in the population, the greater the value of L.*

The match-binning approach does not make full use of d and x. "Pure" match-binning merely asks whether d is inside or outside the predefined match window. This offers a simpler, but cruder represe of the "data" in (1). The fragments either match or they do not - there are no shades of gray:

> *This likelihood ratio is relatively easy to compute. Suppose, as in the example of Part II, that there is a match with a bin frequency of P=.1 The denominator is the probability of declaring a match when -S is true. When the suspect and the source of the crime scene DNA are unrelated, this probability cannot exceed a generous estimate of t proportion of the Caucasian population with a matching band.*
>
> *The numerator is the probability of a match given S. It depends on the size of the match window. It approaches one (1) as the match window grows very large. The FBI, for example, uses a match window that is slightly larger than the biggest difference observed in its laboratory for the same sample measured twice. Some critics have complained that this window amounts to plus-or-minus four standard errors of measurement. This, they say, is too large. Although figure of four standard errors may be misleading, there are indications that the errors follow the ubiquitous normal curve. Since 99.994% of the area under a normal curve lies within four standard errors to either side of its center, an examiner who always declares a match for a band when the band from the crime sample and the corresponding band from the defendant's sample are within the [+ or -]2.5% window will be correct in 99.994% of the cases when S true: Pr(m=1[vertical bar]S) = .99994. More generally, if A(w) sta the area within a match window that extends a distance [+ or -] w about the center of the normal error curve for duplicate measurements, the match-binning likelihood ratio becomes:*
>
> $$L = A(w)/P.$$

Two salient points emerge. First, there is some tension between the similarity likelihood ratio (2) and the match-binning ratio (5). The former attends to the precise degree of matching; the latter pretends that all "matches" are equally revealing. In a case in which the value obtained from (2) is much less than that obtained with (3), the match-binning likelihood ratio can give an inflated impression of probative value.

Second, even within the confines of (5), P is incomplete because it relates solely to the denominator of L. A juror who implicitly thin in terms of likelihood ratios, and who is not apprised of A(w), might treat A(w) as if its value were 1, and weigh the evidence as if its probative value were given by the following "P only" likelihood ratio:

$$L = 1/P.$$

Comparing (5) and (6), if A(w) were dramatically less than 1, then presenting only the value of P could lead a jury to give the fact of a match more weight than it deserves. If this is the argument for exclude match-binning evidence, however, it has little force. As Part IV explains, the match windows are wide enough to ensure that matches are probative.

## Thompson's Argument

Professor Thompson agrees that in many cases the demand for narrower windows would not assist the jury. His argument for admitting testimony about the width of the match window is terse. He cautiously suggests that it "might well be appropriate" to exclude testimony of a match "if the defendant makes a strong showing prior to trial that the observed discrepancies are unlikely to have arisen i the `matching' prints have a common source." This showing, he thinks, would support a defense argument that "the match criteria should have been drawn in a narrower manner." Noting the inability of "lawyers and jurors to deal with complex scientific and statistical arguments," Thompson concludes that Rule 403 "could properly" justify exclusion of the evidence. But if testimony as to a match be admitted, he insists that reasonable people could not differ on the relevance of testimony favoring a smaller window that would not produce a match.

His argument, however, is not conclusive. In more detail, the argument splits into two branches, as follows:

- I When the numerator of the appropriate likelihood ratio is substantially less than 1, P is a misleading measure of the probative value of a match.
- IIa Expert testimony that a smaller window would exclude the defendant might prevent the jury from being misled by the discrepancy between 1/P and the more appropriate likelihood ratio.
- IIIa Therefore, courts should allow expert testimony that a smaller match window would exclude the defendant.

IIb Alternatively, explaining to a jury that the numerator of the appropriate likelihood ratio is substantially less than 1 would be too hard or ineffective.

IIIb Therefore, courts should exclude evidence of the match and of the match frequency P.

This formulation discloses an ambiguity. Which likelihood ratio - (2) or (5) - must be substantially less than I? If (I) refers to (5) then for the hypothetical case outlined in Part II, the likelihood rat is L = .99994/.12 = 8.327. A juror who implicitly uses the over simplif "P only" ratio (6) would think that L = 1/.12 = 8.333 - hardly a mistake that cries out for redress. This is because the condition in (that might produce a misleading picture is not satisfied - the numerat is not substantially less than 1. Instead, it is all but identical to

> *Suppose, however, that the condition in (I) were fulfilled - that a laboratory used a window amounting to [+ or -] 1 standard errors inste [+ or -] 4. Then the numerator A(w) becomes .6827, a value that is substantially less than 1. It follows that L = .6827/.12 = 5.689 instead of the "P only" value of 8.333. In this event, the argument still fails, bec is false. Expert testimony that a smaller window would exculpate the defendant would not prevent the jury from being misled. The expert concerned with the match-binning likelihood ratio would have to testify that the laboratory should use a larger window - exactly th opposite of the testimony that has been heard to date.*

So if the focus is on match-binning proportions, the argument founders at the outset. If Professor Thompson was writing about the match-binning likelihood ratio, he would be correct in identifying PR(m[vertical bar]S) as a possible source of cognitive error. Instead justifying smaller match windows that would exonerate the defendant, however, a small Pr(m[vertical bar]S) would motivate larger windows that would to inculpate the defendant.

Thus, Thompson must not be concerned with the match-binning likelihood ratio. Instead, his argument may be that when the numerator Pr(m[vertical bar]S) of the similarity likelihood ratio (2) is "quite small"[56] P is misleading and either justifies testimony that the match window is too large or warrants exclusion of the match altogether.

But even under this interpretation, what is in question is the full likelihood ratio, not just its numerator, and the event of an observed distance d between two bands is mathematically very different from the existence or non-existence of a match. Whereas the numerator of the match-binning likelihood ratio may be calculated as an area A(w) under a part of a normal curve, the numerator of the similarity likeli ratio is more like the area of a thin strip under the curve located a distance d from the peak. Being a thin strip under a low-lying curve rather than a broad region under the entire match window [+ or - the numerator Pr(x,d[vertical bar]S) always will be small. Likewis the denominator in (2) is not the proportion P of matches within a fixed, overly broad window, as it was in (5). Now, the denominator is Pr(x,d[vertica the probability (actually, the probability density) of seeing two band that represent the same DNA fragment separated by d and centered at x, and it changes as x or d changes. just as the numerator can be much smaller than 1, this denominator can be much smaller than P. P, after all, is the proportion of matches estimated with a bin that i bigger than the match window, which, in turn, is bigger than the small difference d. Pr(x,d[vertical bar]S) will involve but a small slice of big bin.

Consequently, the "substantially less than 1" or "quite small" condi of (I) is neither necessary nor sufficient to establish that the "P only" likelihood ratio 1/P is too large. In principle at least, the nu can be much less than 1, but the probability of seeing the same degree of matching with an innocent suspect also may be much less than P. Therefore, a small numerator in (2) does not establish that (2) is substantially less than (5). In short, it is not obvious t Thompson's criterion accurately picks out those cases that could invol the cognitive error at issue. Once again, the argument for expert testimony about the desirable size of match windows falters at the first step.

## Reformulating the Argument and Solving the Problem

When inspected, Professor Thompson's argument about match windows is less than airtight. Nevertheless, the underlying concern that P is an imprecise proxy for probative value could be valid. Part explained that the match-binning likelihood ratio pretends that all matches within an arbitrary window are equally revealing. Sometimes the more precise similarity likelihood ratio (2) exceeds the match-bin ratios. Other times, the match-binning ratios exceed the similarity likelihood ratio. If the disparity is serious, relying solely on P as measure of probative value could be misleading. But instead of identified these cases by looking solely to Pr(x,d[vertical bar]S), as Thompson recommends, it is better to look to suitable computations of the similarity likelihood ratio. This captures the spirit of Thompson's analysis and avoids the initial misstep.

If this reformulation is correct, the pivotal question becomes what to do about the possible discrepancies between the best measure of probative value (2) and the measure (6) that currently is put befor a jury. Clearly, the correct response is not testimony about the stati properties or "scientific" acceptability of particular match windows. The objection, at its root, is to the entire enterprise of binary matching.

The most radical reform would be a rule that barred prosecution testimony about match-binning proportions, but admitted testimony about likelihood ratios. While prosecutors have been advised to rely on the similarity likelihood ratio instead of P,[61] unless the match-ratios are typically far from the mark, compelling them to do so may amount to overkill.[62] The American criminal justice system gener relies on the parties to pick and choose from the available evidence and to build the case that they think is best. When the value of the match-binning ratio exceeds that of the similarity likelihood rati but the prosecution prefers to rely on the simpler and perhaps more easily comprehended approach, the defense can hardly object that the prosecution has understated the strength of its own evidence.

In contrast, a court might permit the prosecution to introduce P, and leave it to the defense to argue that the better measure of probat value paints a different picture. Such adherence to the usual system of "free proof" - the analog of the economist's "free market" - is reasonable as long as two conditions hold. First, the identification o the cases in which P is misleading must not be prohibitively difficult This consideration counsels requiring the government to disclose to the defense in advance of trial its best estimate of the similarity li ratio.

Second, the free proof system would fail if jurors would be unduly influenced by P even when the defense counters it with cross-examination and testimony about the similarity likelihood ratio. If reasonable defense efforts would not prompt the jury to appreciate the prosecutor fallacy in relying on P, then the prosecution does not risk being shown to be acting unfairly by intentionally or unwittingly presenting evidence that overstates its case. Professor Thompson implies that the type of "market failure" is a distinct possibility. He characterizes t effort to present the similarity likelihood ratio as "a formidable that "arguably would impose an unfair burden on a criminal defendant." The point is indeed arguable, since it is not clear how juries would react to a combination of P and the similarity likelihood ratio. The general philosophy of most modern evidence codes is to allow the jury to hear the evidence unless it is clear that prejudicial effect outweighs probative value, but few appellate courts would find an abuse of discretion it a trial judge limited the jury to receiving the expert description of import of the DNA profiles in the form of a likelihood ratio.

Other reforms lie between these extremes. The forensic scientist could include the likelihood ratio in a written report accessible to t prosecution and the defense. If the ratio is markedly less than 1/P, the prosecution could be barred from introducing P without contemporan testimony during the direct examination of a prosecution expert to the following effect: (a) it would be wrong to rely on P also in weighing the DNA evidence; (b) it is important to compare the chance that comparable DNA profiles would arise when the defendant is the source to the chance when the defendant is not the source; and (c) this comparison shows that in this instance P exaggerates the worth of the DNA evidence. This policy of full disclosure reinforce the norm of fair use of the evidence by the government without deprivation the prosecution of the opportunity to place an appropriately computed value of P before the jury.

Ultimately, even this solution seems unsatisfactory. If clear and contemporaneous testimony about the similarity likelihood ratio success in ensuring that the limitations on P are explained in an effective way, what is the point of the initial testimony about P? If the jury does not rely on P, then the time and argument expended on it is a waste of resources - in economic terms, a dead weight loss. On the other hand, if the jury does use P to gauge the probative value of the DNA profiles, it makes a potentially serious cognitive error; by defining the case is one in which 1/P is much higher than the similarity ratio (2).

In this way, the result, in broad outline, might resemble Professor Thompson's. In a subset of cases match-binning proportions should be excluded, but the DNA profile evidence should be admissible when explained in terms of a suitable likelihood ratio. The subset of c may differ from those identified by Professor Thompson, but it consist of precisely those in which the form of prejudice he identifies is a serious risk. The response to this risk is neither outright exclusion DNA evidence nor admission of nonresponsive testimony about the properties of match Windows. If the problem is sufficiently widesp the best solution may be to use the similarity likelihood ratio to more fairly present the DNA data. ("`Relevant evidence' means evidence having tendency to make the existence of any fact that is of consequence to the determination the action more probable or less probable than it would be without the evidence." Evid. 402 ("All relevant evidence is admissible, except as otherwise provided... Fed. R. Evid. 403 ("Although relevant, evidence may be excluded if its probative value i substantially outweighed by the danger of unfair prejudice, confusion of the issues, or mislead jury, or by considerations of undue delay, waste of time, or needless presen cumulative evidence."). (2) The DNA at the crime scene may come from the defendant's blood, se saliva. In other cases, the victim's DNA may be found on the defendant's prope e.g., State v. Bible, 858 P.2d 1152. Although the discussion here explicitly refers to only the former situ analysis applies to both methods of linking a defendant to a crime. (3) Some courts seem to require such testimony. E.g., Nelson v. State, 69, 76 (finding trial court's exclusion of match frequency "inher inconsistent" with its admission of testimony of a match, because "without the neces statistical calculations, the evidence of the match was `meaningless' to the jury."); State v. B N.W.2d 30, 31-33 (ruling expert testimony that "the likeli person matching in all four fragments...would be one in several billion" admissible, since "without statistical evidence, the ultimate results of DNA testing w become a matter of speculation."); State v. Vandebogart, 616 A.2d 483, 494 virtually meaningless without a statistical probability expressing the frequency which a match could occur."). It would not, however, be "meaningless" to inform that two samples match and that this match makes it more probable, in an amount not precisely known, that the DNA in the samples comes from the same person when all estimates of the frequency are in the millionths or billionths, it wou meaningless to inform the jury that there is a match that is known to be extremely the general population. D.H. Kaye, The Forensic Debut of the National Research Cou Report: Population Structure, Ceiling Frequencies, and the Need for Numbers, 34 Jurimetrics 380-82 (1994). (4) United States v. Jakobetz, 747 F. Supp. 250, 263 (D. Vt. 1990) ("substructure is arguably the weakest link of the DNA profiling chain"), aff'd, 955 F.2d 786 (2d cert. denied, 113 U.S. 104 (1992). See also Committee on DNA Technology in Forensic Nat'l Research Council, DNA Technology in Forensic Science 79 (1992) [herein Report! ("whether actual populations have significant substructure for used" is "the key question"). "Substructure" refers to the presence of subgroup have distinctive DNA patterns and that tend to mate among themselves. For the view that sometimes have erred in excluding the estimates on this ground, see D.H. Kaye, D Evidence: Probability, Population Genetics, and the Courts, 7 Harv.J.L. & Tech. (1993). (5) See Kaye, supra note 4; William C. Thompson, Evaluating the Admiss New Genetic Identification Tests: Lessons from the "DNA War," 84 J. Crim. L. & Cri 22, 50-51 (1993) ("Only one court has held that the dispute over the FBI's match procedure precludes the admissibility of the test under Frye."). (6) Thompson, supra note 5. (7) See, e.g., People v. Simpson, No. BA097211. (8) Thompson, supra note 5, at 60. Professor Thompson recommends exclu when "the defendant makes a strong showing prior to trial that the obs discrepancies are unlikely to have arisen if the `matching' prints have a common sou (9) Id. at 60 n.171. Kaye, supra note 4, at 114-15 (footnotes omitted). Although this theory has deep roots in the works of logicians, philosophers, psychologists, and statisticians, much of the exposition and development in the legal literature has come from the University of Michigan Law School. See, e.g., Richard Le New Evidence Scholarship: Analyzing the Process of Proof, 66 B.U. L. Rev., reprinted in Probability and Inference in the Law of Evidence: The Uses and Limits Bayesianism 61 [hereinafter Tillis & Gr Richard Friedman, Character Impeachment: Psycho-Bayesian []?] Analysis and a Proposed Ov UCLA L. Rev. 637 (1991); Richard D. Friedman, A Close Look at Probativ 66 B.U. L. Rev. 733 (1986). The "Michigan School," however, emphasizes Bayes' rul combining new information with prior probabilities of hypotheses. The approach u this Article focuses on the concept of relative likelihood which can be embraced wi commitment to

Bayesian reasoning in general. See, e.g., D.H. Kaye, Introduction: Bayesianism?, in Tillers & Green, supra at 1. There are critics of the Michigar Scho they do not always offer clear alternatives. See, e.g., Ronald J. Allen, Factu Ambiguity and a Theory of Evidence, 88 Nw. U. L. Rev. 604, 627-28 (1994) (sketching a "theory of evidence" that "reduces to the proposition that a disinterested fact finder reco the past based on all the observational inputs available at the moment of judging"). A chromosome is a tightly coiled molecule of DNA packed in protei parent supplies one of 23 different chromosomes, so the human genome consists chromosomes arranged into 23 pairs. For present purposes, DNA may be thought of as a book whose lette "base pairs." Bacterial enzymes are used to cut the DNA "text" into fragment "restriction enzyme" binds to DNA when it encounters a certain short sequence of DN pairs and cleaves the DNA at a specific site. For example, the Hae III enzym like a scissors cutting the text...GGCC...GGCC...to yield the fragment CC...GG. "Digesting" DNA with such an enzyme usually produces fragments ranging from several hu several thousand base pairs in length. The molecular weight of a compound is equal to the total mass of constituent atoms. Since DNA fragments all have pretty much the same mix of atoms, fragment that has twice the length of another also has about twice the molecular wei In many instances, the bands are sharp and clear; sometimes, they fuzzy. Indeed, in some cases there are disputes over the presence or absence of bands laboratory artifacts sometimes produce missing bands or spurious bands. See David Balding & Peter Donnelly, How Convincing is DNA Evidence?, 368 Nature 285, 286 (1994); William C. Thompson & Simon Ford, The Meaning of a Match: Sources of Ambiguity in the Interpretation of DNA Prints, in Forensic DNA Technology 93, 138 (M.A. Farley & J.J. Har eds., 1991). This Article limits itself to the problem of determining the size of t window. To keep the focus sharply on this issue, it considers only cases in wh is no doubt as to the existence and measurements of the bands on the autoradiogram. W are present, uncertainties created by missing or extra bands and deviation objective procedures for ascertaining the positions of bands can degrade the probative valu declared match. Cf. D.H. Kaye, Comment: Uncertainty in DNA Profile Evidence, 6 Sci. 196, 196-97 (1991) (noting the potential importance of these factors). Bruce Budowle et al., Fixed-Bin Analysis for Statistical Evaluati Continuous Distributions of Allelic Data from

VNTR Loci, for Use in Forensic Comparisons, 48 AmGenetics 841 (1991). Since epithelial cells from the swabs and blood cells from person contain identical DNA, the measurements would b the same in the absen measurement error. An alternative corresponding to rape cases would be to compare blood samples from the sane man. B.S. Weir & B.S. Gaut, Matching and Binning DNA Fragments in Fore Science, 34 Jurimetrics J. 9 (1993). Budowle et al., supra note 16, at 844. Because the speed at which fragments move through the gel varies different regions of the gel, the absolute size of the experimental error tends larger for larger bands. Hence, the window is a percentage of the molecular weight or si fragment. In light of the studies cited above, the FBI uses a window of [+ or -] the South Carolina laboratory uses [+ or -] 2.8%. For the purposes of this Article, a false negative is a failure t a match between two samples of DNA from the same person. One group of researchers foun match window of [+ or -] 1.2% with three probes produced false negatives nearly 20% of the time. The rates would be larger for four or five probes, since each pr creates more opportunities for a false negative to occur. I.W. Evett et al., An Ill of the Advantages of Efficient Statistical Methods for RFLP Analysis in Forensic Science J. Hum. Genetics 498, 502 (1993). For present purposes, a false positive is a declaration of a matc DNA from different people. It is possible to argue that some minimum values of the sensitivi specificity of the test are needed to establish scientific validity. The need for sci validity (under the rubric of "evidentiary reliability") is emphasized in Daubert v. M Pharmaceuticals, Inc., 113 S. Ct. 2786 (1993). The first two federal a cases applying the Daubert standard of "good [scientific] grounds" to DNA evidence ar States v. Bonds, 12 F.3d 540 (affirming the admission of a ma record developed in the extensive pre-trial hearing in United States v. Yee, 161 (N.D. Ohio 1991)), and United States v. Chischilly, 30 F.3d 1144 (9th Cir. 1 (affirming the admission of a match and a "conservative" estimate of the genotype fre among Navajos). The need to demonstrate scientific validity under Daubert may require reproducibility studies to establish that a laboratory is capable of declaring a match samples from the same person. See Thompson, supra note 5, at 56-57. But the minimum rates that make a clinical test acceptable to scientists (or to anyone else) depe costs of false positives and false negatives in light of the uses to which the test i specificity, for example, may be acceptable in a test, like a pap smear, that is used t for certain conditions whose presence can be confirmed by more definitive (and mor expensive) tests. Rather than ask the abstract and unanswerable question of what sensitivity and specificity science demands, it seems more profitable for courts t directly into the helpfulness of the test to the jury. In other words, the "science" requirement of Daubert should apply to the determination of whether the analytic procedure is of performing as warranted by scientists and technicians - whether it is indeed scientific - but not to whether the advertised or known level of performance is satisfa legal purposes. See Kaye, supra note 4, at 114-15 n.67; Kathryn Roeder, Rejoinder Sci. 267, 275 (1994) ("the objective' match criterion...is, in fact, simply an arbi rule."); see also infra note 43. See supra note 4. Often, P is interpreted as the probability that a randomly select innocent person will have DNA that matches the crime scene sample. If so, it is akin t statisticians usually call a "p-value" or "significance probability" - the probability the "null hypothesis" of observing data within a region (the match window) that (when small prompts rejection of the "null hypothesis." Under the null hypothesis someone other than the defendant is the source of the crime scene DNA (and assuming suspects are not selected because of factors related to their DNA profiles and that laboratory errors occur), the probability of declaring a match is just the proportion of the population who have profiles that satisfy the match criteria. See United States v. Jakobetz, 955 F.2d 786 (2d Cir.), cert. deni U.S. 104 (1992). Computing P when more than one band is compared in the two sample proved intensely controversial. For reviews, see, for example, Kaye, supra no Kathryn Roeder, DNA Fingerprinting. A Review of the Controversy, 9 Stat. Science. 222 (199 Thompson, supra note 5. The issue here, however, is not the details of combining the i from the analyses at each band, but what use to make of even undeniably accurate estimates of P, either for a single band or for many bands. For convenience, the text considers the simplified, one-band comparison. A variation uses "floating" bins. See, e.g., Keith L. Monson & Br Budowle, A Comparison of the Fixed Bin Method with the Floating Bin and Direct Count Methods of VNTR Profile Frequency Estimation and Reference Population, 38 J. Forensic; B.S. Weir, Independent of VNTR Alleles Defined as Floating Bins, 51 Am. J. Hu Genetics 992 (1992). Because the differences do not affect the results of the analy the discussion is limited to fixed bins. The choice of ten bins is illustrative; in practice, more may be See, e.g., john S. Waye & Ronald M. Fourney, Forensic DNA Typing of Highly Polymo Loci, in 3 Forensic Science Handbook 358, 385. T values of P that underestimate the proportion of matches that would be obtained population, the width of the bins must be such that no bin is smaller than the mat In practice, considerably larger bins often are used, arguably leading to P that are too large and thus understate the rarity of a match. But see Thompson, note 5, at 67 (reporting that only Cellmark Diagnostics uses bins larger than its "m standards"). If the crime sample band were close to the boundary of two bins, ordinarily would be safe to use the larger of the two bin heights - 12% in this e extremely cautious approach would be to use the sum of the heights - 12 + 8 = 20 NRC Report, supra note 4. See supra note 11. The hypothesis -S can be thought of as the disjunction of hypothe everyone else in the relevant population. If all the people who might be the so the crime scene DNA are labelled 1 through n (with the defendant being designate person 1), and if the defendant, who is the source of the DNA being compared to t scene DNA, is not the source of the latter sample, then either person 2 is t person 3 is, and so on, through person n. See A.W.F. Edwards, Likelihood: An Account of the Statistical Con Likelihood and its Application to Scientific Inference (1972). See, e.g., Richard O. Lempert, Modeling Relevance, 75 Mich. L. Re (1977). For other possible expressions of probative value, see Friedman, supra not Kaye, Comment: Quantifying Probative Value, 66 B.U. L. Rev. 761 (1986). See Thompson, supra note 5, at 58 (figure 3). On computing L, see David Jarjoura et al., Likelihood Ratios for Deoxyribonuleic Acid Typing in Criminal Cases, 39 J. Forensic Sci. 64 (1994); Kaye, s note 4, at 162 n.254; Roeder, supra note 27. While grossly oversimplified, this expos intended to capture the features of L that are important to the analysis of a clai match window is too broad. In more technical jargon, instead of treating d as the continuous that it is, match-binning converts d into a binary variable, throwing away informal practice, however, the declaration of a match is not so black-and-white. For one human judgment may enter into the determination of whether a faint band actually is p whether seemingly extra bands are laboratory artifacts. See supra note Furthermore, even if the full set of bands to consider is clear, and even if they m statistical criteria for a match, subjective judgment may be involved in deciding whether t bands "really" match. Realizing that an electrophoretic artifact known as band-shifti sometimes results in all the bands being shifted in more or less the same proportion fro true values, an examiner may decide that two profiles match when all the difference between the samples are in the same direction and close to the edge of the match w but not when the differences go in both directions. The former situation is co with a very close match distorted by band-shifting; the latter is not. To the exte examiners rely on the actual size of the differences and not just the fact that the differences fall within the preset window in declaring matches, the match-binning procedure does u of the information about the degree of matching.

It is difficult to model the effect of these subjective aspects on p value. For the purpose of analyzing the relevance of testimony about match windows, t Article assumes that there are no ambiguities in the existence of bands. As for the se subjective phase of matching, since examiners presumably screen out the cases in which are weak more often than they discard excellent matches, one might argue that t review can only enhance the probative value of a purely statistical match. Even i subjective judgments were entirely arbitrary made, for example, by flipping a coin to dec whether a statistically tolerable match will be deemed a match the subjective still could not lower the probative value. Consequently, the type of match-binning described and analyzed in this Article - involving purely statistical matching - may understate the probative value of those matches that arise in practice. But see infra note 46. other hand, if Professor Thompson is correct in his claims that examiners frequent declare matches that do not satisfy statistical criteria, then the analysis applicable purely statistical matching could overstate the probative value of the matches that arise in pract But cf. David J. Balding et al., Comment. Some Causes for Concern DNA Profiles, 9 Stat. Sci. 248, 249 (1994) (One needs to consider each possible cul to assess the probability that they would match the crime profile, given that th defendant does."); B.S. Weir, Conditional Genotypic Frequencies in Forensic Analysis (Oct 1993) (unpublished manuscript focusing on "[t]he conditional probability that the accused suspect has the matching profile, given that die perpetrator of the crime has the and that these people are not the same"). See supra note 29. Balding et al., supra note 38, at 249, contend that one must cons likelihoods of each alternative hypothesis in note 32 supra, and that "statistic analyses of forensic databases are not directly relevant to an assessment of the conditional probabilities" that determine these likelihoods. They argue that the very fact that the de matches makes it more probable that relatives of the defendant and others in t ethnic subpopulation also would match. Because of these "positive correlation conclude that using the population proportion is "both false and detrimental to defence." Id So too, Balding and Nichols maintain that the conditional probability from the estimated population proportion because "if the suspect is innocent have observed two copies of a profile which is believed to be extremely rare in the population," and "this observation makes more likely the possibility that the profi more common in the suspect population than in the database population." Dav J. Balding & Richard A. Nichols, DNA Profile Match Probability Calculation, 64 Fore Int'l 125, 128 (1994). Thus, these researchers propose using the proportions for the defendant's ethnic subpopulation rather than the broad racial population in estimating th likelihoods. Balding et al., supra note 38, at 249. See supra note 19. It also requires the examiner to say that profiles that meet this look like matches. E.g., Seymour Geisser, Some Statistical Issues in Medicine and Fo 87 J. Am. Stat. Ass'n 607, 609 (1992). The standard error is the standard deviation of difference between two measurements; it indicates the variability of the different a sample batch of differences, it is computed by subtracting each difference fr mean difference, squaring this deviation from the mean, adding up all the s dividing by the size of die sample, and extracting the square root. In other words the root mean square deviation. If all the differences are identical, the deviations mean are all zero, and the standard error will be zero. If all lie at two polar ext all the deviations are large, and the standard error will be large.

The standard error of DNA molecular weight measurements is not reall constant percentage of the true weight. See I.W. Evett et al., An Efficient Sta Procedure for Interpreting DNA Single-Locus Profiling Data in Crime Cases, 32 J. For Sci. Soc'y 307, 316 (1992). In the interest of brevity, this Article ignores thi complication. Geisser, supra note 42 ("an extraordinarily wide net"). Some expe United States v. Yee, 134 F.R.D. 161 (N.D: Ohio 1991), aff'd sub nom., United States 12 F.3d 540 (6th Cir. 1993), claimed that the FBI's window is too wide because "the of risk for false positives (matches)" exceeds that which is generally accepted a reliable in the research and clinical communities for similar applications of this technology." Thompson, supra note 5, at 48-49 nn.122-23 (quoting Report of Professor Peter D'Eustachio, An Evaluation of the FBI's Environmental Insult Validation Study and the Quantitative Matching Criteria, at 16, in Yee, 129 F.R.D. 629). Since these communities have similar applications of the technology, however, this version of the criticism odd. Unlike a diagnostician or (perhaps) a research scientist, the forensic expert n and should not choose between two hypotheses. Rather, the forensic expert faces t challenge of presenting the scientific evidence bearing on the hypothe that the jury will appreciate its actual probative value. The [+ or -] 4 figure pertains to the standard error derived from cell line (control DNA) measurements. These involve fresh DNA that produces less variable measurements than forensic samples, reducing this size of the standard error, and m appear that a given difference represents a larger number of standard errors. Thus, S. Lander, Invited Editorial Research on DNA Typing Catching Up with Courtroom Application, 48 Am. J. Hum. Genetics 819, 820 (1991), reports that the Bureau's laboratory ha standard error of about 1.5% of the molecular weight of their mean. If so, the [+ or match window corresponds to only [+ or -] 1.7 standard errors. Donald A. Berry et al., Statistical Inference in Crime Investigation. Using Deoxyribonucleic Acid Profiling, 41 Applied Stat. 499 (1992). This applies to a single band. For most people, eight or more ban compared According to Bonferroni's inequality, the probability of at least one in eight comparisons is, at most, 8 x .00006 = .00048. Therefore, the probability eight matches cannot be any less than 1 - .00048 = .99952. Examiners who rely on the impressions of band-shifting to exclude some statistical matches, see supra note 37, lower the probability of declaring a match when the DNA samples originate from t individual. The results of matching with a subjective element could have a li ratio closer to that given by (2). See supra note 37. If A(w) were less than P then the likelihood ratio (5) would indi the evidence does not support the hypothesis of a common source. Cf. Thompson, supr at 57. To some extent, this reinforces the conclusion of Part I. Unles laboratory can demonstrate that it can reliably declare matches for DNA from the same the evidence should be inadmissible. However, in terms of logical relevance as oppo the more demanding Daubert standard, what counts is not whether the chance declaring a match when S holds close to 1, but whether it is substantially larger chance under -S. If the DNA bands are extremely rare, then the latter quantity can small, so that even modestly reliable DNA tests may produce highly probative matches. Thompson, supra note 5, at 60. Id. Id at 60 n.171. See also id. at n.170. Id. at 60 & n.171. Id. at 60 n.171. This assumes that the number of bins remains unchanged. With a sm match window, more bins with smaller bin widths would be appropriate. This c however, could only reduce P (since fewer bands would fall into most of the sma bins). Reducing P would raise both the correct value of the match-binning likelihood r A(w)/P, and the perceived value, 1/P. The illustrative numbers would change, but t general picture would not. See United States v. Yee, 134 F.R.D. 161 (N.D. Ohio 1991), aff'd United States v. Bonds, 12 F.3d 540 (6th Cir. 1993); supra text accompanying Thompson writes: "In formal terms, the defendant should make a sh that p(D/H1) is quite small." Thompson, supra note 5, at 60 n.168. The expression p in the notation used here, would be Pr(d[vertical bar]S). The text of his art imposes the condition in (I) that "the observed discrepancies are unlikely to have arisen if the `m prints have a common source." Id at 60 (emphasis added). The "observed discrepancies abbreviated as d in equation (2) and D in Thompson's expression for the difference likelihood ratio. Id at 57 n.163. His definition of the likelihood ratio does not explicitly to the location of the "matching" bands, presumably to make die expression ea read. This "thin strip" description is an approximation. Because d is a continuous variable, (2) is the ratio of probability densities rather than probabilities, a numerator is given by the height of the normal curve at d rather than an area beneath the Even two bands that are very close will produce a small value of Pr(x,d[vertical bar]S). For instance, discrepancies of just about one standard error occur well under 1/3 of for normally distributed errors. One third is the approximate probability discrepancies of one or more standard errors. The quantity in the similarity likelihood (2) is not this tail-end probability, but rather the probability density for exactly o standard error. See supra note 57. This density might be given by the height of the nomina at one standard error, which is 0.24. The height of the curve for two perfectly aligne - the best result one could hope for when the samples are from the sam - is less than 0.40.

Of course, the normal curve is merely a model that fits the observed distribution of errors in repeated measurements. A particular reproducibility study mi that more than 40% of the replicates are perfectly aligned, or that fewer are pe aligned. One study using one version of the difference likelihood ratio wi on three of Lifecodes' probes reports that, on average, the combination of bands a the three loci produce monotonically increasing ratios with decreasing values of study finds that at a separation of three standard errors, one probe gives an aver likelihood ratio below 1 (supportive of -S). At four standard errors, the average ratio probe is much less than 1. Jajoura et al., supra note 36, at 70 (table 1). The to which this study indicates that focusing on the numerator of the difference likel ratio might identify the troublesome cases is unclear. Being averages, these value pertain to all cases. Moreover, they are averages for a ratio rather than just the nu If Professor Thompson wished to test his criterion empirically, he could identify t cases in which the difference likelihood ratio is substantially less than 1/P in a da obtained by measuring many fragments twice and then see how well these discrepancies correspondence "quite small" values for Pr(x,d[vertical bar]S). Of course, actual practice departs from the model of purely binar matching. See supra note 37. See, e.g., Donald A. Berry, DNA, Statistics, and the Simpson Case Chance, Fall 1994, at 9, 10; Donald A. Berry, Inferences Using DNA Profiling in Forensic Identi and Paternity Cases, 6 Stat. Sci. 175 (1991); J. Buckleton et al., A Continuos Model Interpreting the Positions of bands in DNA Locus-Specific Work, 31 J. Forensic Sci. Soc (1991); Bernard Devlin et al., Forensic Inference from DNA Fingerprints, 87 J. Am. Sta 337 (1992); Evett et al., supra note 42. Evett favors characterizing the numerical value descriptions like "very strong evidence of identity." Id. Jarjoura et al., supra note 36 propose that the expert also provides the jury with the proportion of times that likelihood ra large or larger than the one at bar have been obtained from different persons i databases. According to Evett et al., supra note 42, at 325 (Table 7), likelihood exceeding 1000 almost always arise for repeated measurements of the same DNA and almo occur in comparing DNA from different people in a database. See also I.W. Evett Statistics: Putting the Problems Into Perspective, 33 Jurimetrics J. 139 (1992) (exhaustive pairing of 1500 Caucasians tested at three loci, generating over one million between-p comparisons intended to simulate cases of false accusations, gave only nine likeli ratios greater than 1, and most of these were below 100). See Evett et al., supra note 42, at 325 (Table 7). See Paul C. Giannelli, Criminal Discovery, Scientific Evidence, a 44 Vand. L. Rev. 791 (1991) (emphasizing the importance of discoverable and meaningful reports). Thompson, supra note 5, at 60. More recently, he has su that likelihood ratios should not be used unless they "accurately take into account" uncertai the examiner's decisions as to whether a band is present and where it lies gel. William C. Thompson, Comment, 9 Stat. Sci. 263, 264 (1994). Even if this view accepted, it would not apply in the category of cases that is the subject of this Article. See supra note 15. One might buttress the argument slightly by recognizing that if t specific likelihood ratio (2) is going to be presented anyway, the jury learns little or n useful from P. What makes (2) superior to (5) and (6), after all, is that it makes more of the data. Other limitations on P that might justify similar, cautionary tes the interest of full disclosure are described in Kaye, supra note 4, and Thompson, note 5. For discussion of how a likelihood ratio might be presented or us Berry, supra note 61; Kaye, supra note 4; Richard Lempert, Comment. Theory and Prac DNA Fingerprinting, 9 Stat. Sci. 255, 258 (1994); Roeder, supra note 27; supra note 61. Whether any special rules are required to handle the subset of ca which match-binning frequencies are misleading is not clear. Without an empirical analysis rate at which the similarity likelihood ratio (2) is substantially less than match-binning likelihood ratios, the size of this subset - and hence the suggestion that there widespread problem that warrants a procedural solution - is speculative. Before d that all laboratories compute similarity likelihood ratios to isolate the cases there would be a discrepancy, empirical analysis to ascertain the extent of the pr would seem advisable.

# 5

# POWER OF DNA

## Introduction

The woman reported the incident to the police. When the police interviewed Dr. Vishal, he denied having had sexual intercourse with his patient. But DNA results using the RFLP method indicated that it was Vishal 's semen found on a vaginal swab taken from the woman at the hospital several hours after her visit to his office. His lawyer challenged the DNA in a pretrial hearing but was unable to keep the judge from ruling that the results would be admitted at trial.

So, at the trial, Dr. Vishal 's lawyer took a different tack, urging that the doctor and his patient had had consensual sex. During cross-examination by the defense lawyer, the woman admitted to being an alcoholic, having a history of drug use, and having received counseling for troubled relationships.

The defense rested without calling Dr. Vishal or any other witnesses in his defense, relying for the most part on impugning the credibility of his accuser, hoping that the jury would not send a medical doctor to prison on the say-so of someone who had been forced to seek help for emotional problems. The Westchester jury acquitted Vishal.

Though Dr. Vishal was thus spared from a prison sentence, this was not the end of the matter. Soon thereafter, the state began civil proceedings to revoke his medical license. After hearing testimony from the patient and Dr. Vishal, the state found that the sexual intercourse between the two had not been consensual and revoked his license to practice medicine.

Had the consent defense worked? It had prevailed in the criminal trial, where the standard of proof was guilt beyond a reasonable doubt, but been rejected in the civil proceeding, where a lesser standard of proof applied. Dr. Vishal had maintained his liberty but lost his license to practice medicine.

A defense based on consent is the one most likely to be asserted in rape cases where DNA testing confirms that intercourse has taken place. As it did in the Vishal case, DNA testing can make virtually indefensible the claim that the suspect did not have sex with his accuser, and the defense must switch to the claim that the sex took place but was consensual. In many cases, that claim will be based on truth. In many others, it will not. But that is not a point DNA can settle.

Oddly, the "frame" defense utilized in the O. J. Simpson trial is in certain legal respects a distant cousin to the "consent" defense in a rape trial. In both cases, a suspect acknowledges the presence of his blood or semen but offers a theory for where it was found that allows for his innocence. The focus shifts from whether it is the defendant's blood or semen at the crime scene to how it got there. Rather than challenging the accuracy of DNA laboratories, this may be the way defense attorneys will deal with the fact that DNA test findings put their client at the scene of the crime. Some of these defenses are a tribute to the ingenuity of defense lawyers, but in the first instance they are a tribute to the power of DNA. It is only because the scientific validity of this means of identification cannot for the most part be directly challenged that realistic defense lawyers have accepted that, if their client is to go home rather than prison after the trial, what is required is an end run around the DNA.

No matter how powerful any new tool, there will always be limits, some never imagined until a bizarre case makes them apparent.

When fifteen-year-old Amanda Hall was raped and murdered in suburban Maryland, it looked like the perfect crime. True, there were witnesses who could link the Page twins to Hall shortly before her murder. The twins had expressed interest in the beautiful runaway, who had been sleeping on the steps of the local recreation center. On the evening that she was murdered, they asked a man who had been talking with her if he was her boyfriend. Later that night, around midnight, they told a clerk at the 7-Eleven across the street that they wanted to buy a hot chocolate for her and were last seen walking toward the recreation center. But there were no witnesses to the rape, or to Amanda Hall's strangulation.

Semen and DNA were recovered from the rape, but that did not resolve the case. The DNA matched both Jerome Page and Tyrone Page. The two are identical twins, and while identical twins do not have identical fingerprints, they do have identical DNA. This left the police stymied. Guilt in the United States is of course an individual matter. A jury cannot convict two brothers because it knows that at least one of the two committed the crime. Since all the DNA evidence could say was that someone with a certain DNA had raped Amanda Hall, and the twins each had that DNA, it alone was insufficient to charge or convict either brother.

But there was a twist to the case, and it lay in one of the ways that identical twins may be different from each other. While identical twins usually have identical hair, the characteristics of their hair can vary if they have had substantially different diets, or if only one has a history of drug or alcohol abuse.

Hall's body had been found the next morning, in a wooded area by some old railroad tracks, two miles from the recreation center. Her body was wrapped in the blanket in which she had slept. Within that blanket, there were pubic hairs that were not Amanda Hall's and were not identical to each other. The FBI analyzed those hairs, comparing them with Jerome Page's and Tyrone Page's. The two brothers' hairs were different from each other, and the hairs of each matched hairs found in the blanket.

One of the twins had previously spent time in jail, doubtless eating a diet substantially different from his brother's. That difference in diet could reasonably have accounted for the difference in the characteristics of the twins' hairs. Whatever the reason, the difference was a fact, and the pubic hairs found on the blanket in which Hall's body was wrapped indicated that both twins had had intimate contact with Hall near the time of her death.

Both brothers were convicted of murder and rape and were sentenced to life in prison without parole. Without DNA, there would have been no case against either twin, since pubic hairs, on their own, are not sufficiently distinctive to form the basis for a criminal conviction. DNA alone was not sufficient to make a case against either brother, because it could not determine if only one or both had done the crime. If the state decided to settle for half a loaf and proceed on the basis of one brother's having done the crime, how could anyone know if it was trying the right brother? The jury verdict that both brothers had participated came out of the totality of the evidence.

The case of the Page brothers, while hardly typical, points out dramatically that DNA does not exist in a vacuum. Other evidence must be contended with. In the Matias Reyes case, the Minnesota data-banking case, and the O. J. Simpson case, there was other evidence, and other issues, that had to be considered alongside the fact of a DNA match. And, in the Central Park jogger case, it was the sum of the evidence that explained the absence of the defendants' DNA.

In most of those cases where there is DNA and some additional evidence against a defendant, that person will choose to plead guilty. This means that victims will be spared the trauma of testimony at trial and that society will be spared the expense and uncertainty of trying guilty people.

But in those cases where a defendant chooses to exercise his right to go to trial, a jury will assess the meaning of the DNA, or the absence of DNA, along with all the other evidence favorable and unfavorable to the person charged with a crime. A person charged with a crime on the basis of DNA tests may have a valid defense based on consent to sex, the planting of evidence, or problems in the collection or handling of the evidence. DNA is a powerful tool for determining identity, but it must always be considered in the context of the individual criminal case.

## Future of DNA Analysis Remains Strong

Scientifically, its capacity to identify the guilty has grown year by year. Over time the RFLP technique has been refined so that it can in many cases analyze more genetic locations than was possible when the technology was introduced. In those cases, it has become more and more like the "fingerprint" it was initially supposed to be, uniquely identifying an individual. This trend can only be expected to continue.

And the underlying soundness of DNA testing was underscored in the aftermath of the Simpson trial when the National Research Council of the National Academy of Sciences issued its long-awaited second report on DNA evidence on May 2, 1996. This second panel had been formed to reconsider the first committee's proposal for limiting DNA numbers, but its report ranged beyond that original mission. The new committee was composed of scientists and law professors mostly different than those who authored the controversial 1992 report. Its essential conclusion: DNA technology and statistics "have progressed to the point where the admissibility of properly collected and analyzed DNA should not be in doubt."

While never mentioning the Simpson case by name, the new report seemed to limit some key avenues of attack that his defense lawyers had happily strolled in challenging the DNA. The new panel expressed views sharply at odds with certain aspects of its predecessor's comments which had proved useful to DNA defense lawyers, although in most instances it did not specifically point out these divergences. Where the previous panel raised the possibility that contamination might not be detected through laboratory controls, the second panel noted that "appropriate safeguards and controls can be built into the analytical system to protect against contamination and to detect it when it does occur." The new committee urged that laboratories "make every effort to be accredited" but, unlike the previous panel, made no suggestion that accreditation be an absolute requirement, now or in the future, for the admissibility or acceptance of work performed by a DNA laboratory. And the new committee scorned the very suggestion (embraced by the first panel) that blind proficiency testing should be used to measure error rates and the implication that such testing should at some future point be mandatory. The proper use of blind proficiency testing, it wrote, is not to measure error rates but to identify problems which need to be corrected and improve laboratory performance, and its utility for even that limited purpose is under review by the National Institute of Justice. The use of such testing to determine error rates would be outlandishly expensive and disruptive, a badly misplaced effort consuming resources much better spent in other ways, such as improving laboratory standards. A more useful protection against laboratory error would be early separation, into two portions, of crime scene evidence, with one reserved for retesting.

Significantly, there were also substantive implications arising from a major shift in tone in the second report. The National Research Council's first report had used sweeping language about its long-range proposals. It could not have gone unnoticed by the new panel that the critics of DNA testing had seized upon that language to argue that the failure to meet those aspirations had rendered DNA testing invalid and inadmissible. So the new panel used more modest words, separated its "recommendations" from its "suggestions," and made specifically clear that it was not making formal recommendations to the effect that its views of matters of legal policy should be prerequisites to the acceptance of DNA evidence. Aware of the potential power of its own words, the committee chose to tread carefully and minimize the risk that its policy preferences might undercut the acceptance of DNA evidence in court.

The new panel also addressed the diminished but once vitriolic controversy over the impact of race and ethnicity on stating the significance of a DNA match, renouncing its predecessor's much-criticized proposal for limiting DNA numbers on practical and statistical grounds, and because there is "no scientific justification for its use." This time, the panel embraced the position advocated by the supporters of DNA testing and endorsed the use of broad racial groupings in stating a DNA match. Where the race of the person who left the evidence is known, a number should be given for that race; where it is unknown, statistics for whites, blacks, Hispanics, and East Asians should be provided. The conclusion that broad racial groupings are acceptable was supported by the new committee's important finding on a previously disputed issue: that "differences between races are considerably larger than those between subgroups within races." According to the panel, there was now "new data" that allowed numerical estimates to be made with greater confidence than before. The frequency of a DNA profile may vary among ethnic groups within broad racial groups as much as ten times in either direction, but this variation simply provides a guide to the level of uncertainty, with the true value estimated to lie between the upper and the lower boundary.

It was clear from the new report that the nation's leading scientists continue to hold the view that DNA testing is fundamentally sound, although issues regarding the collection and analysis of evidence may still arise in individual cases.

The popular view of DNA is another matter. It is certainly possible that a popular misperception that the Simpson trial somehow discredited DNA will lead to the outright acquittal of some defendants squarely implicated by it. But any such effect will surely be short-lived, for DNA is overwhelming evidence of identity, and, in most cases, the courtroom is an empire of reason, where fully credible evidence will continue to have influence.

To the extent that defense lawyers attack the reliability of DNA outright, rather than focus on contextual issues such as consent, planting, or specific problems in the field or laboratory – they will likely fail. They will be like King Canute, ordering the waters to recede. It simply will not happen.

Still, there is a substantial danger that the Simpson case may pose a setback to DNA outside the courtroom that will reverberate ominously throughout the criminal justice system. That will happen if the case results in an increased reluctance on the part of prosecutors to request DNA tests because of a misbegotten belief that such tests are so complex and their explanations so difficult to deliver that they will inevitably confuse jurors and create a fertile field for the planting of reasonable doubt.

Analysis of DNA was relatively infrequent even before the Simpson case. Prosecutors and crime laboratories do not routinely arrange for DNA analysis in all cases where it can be used.

The tests are expensive, and many prosecutors' offices and crime laboratories are not well enough funded to allow such broad application of this tool. Besides, many individual prosecutors are wary of the new world of criminal justice DNA threatens to create. They would prefer to continue doing things the same old comfortable and familiar way they have always done things, relying on the conventional triad of criminal proof and leaving science to those of their college classmates who attended medical school or graduate school in the sciences. Lawyers like to be in charge, and in a courtroom, as in any intellectual arena, the lawyer who has the knowledge has the power. What lawyer would relish getting involved in either an extended hearing or battle at trial about scientific material so complex and esoteric that the lawyer comes off as poorly versed compared with the witnesses he or she must examine?

In this regard as well the Simpson trial may exacerbate the reluctance of prosecutors to introduce DNA experts into their private preserve. Here the experts brought in were so well informed, and of such high repute, that to examine and confront them the prosecution had to borrow better-qualified prosecutors from other jurisdictions. This is not a picture the average prosecutor wants to use as a model.

Often criminal defense lawyers are just as hesitant, if not more hesitant, to push for DNA testing. Hardened by long exposure to career criminals, many defense lawyers start off believing that their clients are guilty, regardless of their clients' protestations that they were not within miles of the scene of the crime on the day in question. As a consequence, many criminal defense attorneys are wary of ordering a test in each individual case that may force them to face that their client has been lying to them and that could potentially be admitted at trial to prove their client's guilt. This presents the criminal defense attorney with a difficult choice: Forgo DNA testing, and risk the conviction of an innocent person, or request such testing and possibly aid the conviction of the lawyer's own client.

So, if the criminal justice system is to deserve its name, prosecutors, crime laboratories, and those who fund their efforts must develop the will, and expend the funds, for routine DNA testing of high quality, for here is a tool that can provide the most accurate information possible on at least one of the issues which dominate criminal trials, namely identification. Developments since the advent of DNA testing have suggested that human inadequacies in discerning innocence and guilt may run deeper than previously thought. It is often said that eyewitness testimony is the most faulty aspect of the criminal justice system, and as long as there are strong-willed police and weak-minded suspects, there will be false confessions. But DNA evidence is virtually unchallengeable on the issue of identity. Unlike eyewitness testimony and confessions, it provides objective evidence of the source of bodily material, which may clear or implicate the accused.

Consider the cases of Colin Pitchfork, Gregory Ritter, and Edward Honaker. In these instances, DNA testing demonstrated once again that human beings do not always know what they think they know. That realization has been underscored by the results of the FBI's DNA-testing program, which has excluded suspects as the sources of samples submitted to it in one-quarter of its cases. This fact is sufficiently noteworthy to bear repetition: In one-quarter of the cases handled by the FBI, the samples tested do not match a suspect's DNA. This does not necessarily mean that one-quarter of criminal suspects are innocent. There may be explanations for many of these test results consistent with guilt. A rape suspect, for example, may not have ejaculated, and the source of semen found in the victim may have been a husband or boyfriend with whom she had consensual sex near the time of a rape. Still, the statistic must give pause, especially when we look at all those cases where DNA testing is not performed.

Few aspects of DNA analysis have been so moving as its use to free people who have spent years in prison after their wrongful conviction of crimes that they did not commit. Yet the number of people so cleared, fewer than twenty overall, is small compared with the number of people routinely cleared before trial by genetic testing. The greatest friend of the falsely accused is DNA testing at the beginning of a criminal investigation; it clears far more people than post-conviction tests. Early genetic testing can spare suspects months and years spent in jail and under suspicion by their communities.

Besides, the days of post-conviction testing are surely numbered. Because DNA testing was not available until recently, legal requests for post-conviction testing were frequently granted on the ground that DNA was newly discovered evidence not available to the defense at the time of the trial. Judges often had no choice but to order the testing of evidence stored for years in police property rooms. By contrast, it is difficult to imagine judges ordering the testing of evidence in cases arising years after the advent of DNA testing, giving convicts who could have requested testing during their first trial but did not a second bite at the apple of an acquittal. Further, justice is best served when the DNA evidence is introduced at the time of the trial, when a rape victim can better remember if there were consensual sexual partners immediately prior to the crime, or what a particular stain in her bedroom might reflect. Timeliness is also a factor when exploring issues that might arise regarding integrity and competence in the collection and storage of evidence.

An expanded commitment to DNA testing should include steps to ensure that new laboratories do not introduce another source of potential error into our criminal justice system. Accreditation or other guarantees of laboratory quality will become even more important as DNA analysis increasingly devolves from the FBI to state and local crime laboratories. Of course, a laboratory may perform good quality work without accreditation. But since many defendants lack money and many defense attorneys lack the expertise to pose a challenge to genetic testing in each case, accreditation, rather than the adversary system, is probably the best way to protect quality for all, rich or poor. It is one thing to set standards but another to make sure they are met, and accreditation is one means to that end.

Because of the enormous weight given to DNA findings, it is important that the laboratories supplying the test results be fully funded, sophisticated facilities, with appropriate outside monitoring. This is of course above all a matter of proper funding and training, but, to protect quality, laboratories should ideally participate in accreditation programs for both RFLP and PCR testing. Today, many do not. The critics of DNA analysis deserve plaudits for highlighting these concerns; measures to ensure the quality of DNA analysis are an important part of promoting the future of the technique.

Analysis of DNA, first used in a criminal case in 1987, is still relatively new. But it is well established as a science. No one maintains that it is "junk science." Questions concerning the fundamental soundness of DNA testing concerning the congressional Office of Technology Assessment are in the words of the, "red herrings" and a "public disservice." Obviously, DNA analysis must be done scrupulously; power, including scientific power, must always be exercised with care, and the greater the power, the greater the care that must accompany its use. An insistence that things be done right hardly renders this powerful technology invalid.

The objective information provided by DNA is among the most reliable evidence available to those seeking truth in our criminal justice system. Criminal convictions and acquittals result from a subjective weighing of evidence presented, and subjectivity is a necessary and proper part of the criminal justice system. But the reliability of the verdicts jurors reach depends on the quality of the evidence to which they are exposed.

There is no question that under the current system the guilty are sometimes set free and the innocent convicted. DNA testing cannot change the system into one where a verdict of guilty or not guilty is preordained by the presence or absence of DNA, for DNA must always be considered in the context of all the evidence. But it can avoid many miscarriages of justice that might occur without it, in a world where the truth is often hidden and elusive. And the use of data banks to solve unsolved crimes can identify repeat offenders guilty of new rapes and help incarcerate them for appropriate periods of time, and so limit their opportunities to inflict more harm and violence on innocent women.

And so, at long last, thousands of years after Cain killed Abel, science has provided the means to hear blood cry out the names of those who have brought violence on others. It is a technology that must be nurtured and protected, for a society that treasures justice cannot afford to turn away from the piercing sound of truth.

# 6

# DNA Computing

## Introduction

Rapid sequencing techniques were introduced a little more than a decade ago, and the rate of sequencing continues to accelerate. As of spring 1991 the international databanks contained approximately 50 × 106 base pairs of DNA sequences with an average sequence length of about 1,000 bp. Improvements in sequencing technology continue to be made, and the associated discoveries in biology are staggering.

It is instructive to look at volumes of Dayhoff 's Atlas and consider the data available in the early 1970s. Hemoglobin and cytochrome c sequences had been determined from a variety of organisms, and efforts were being made to infer the evolutionary relationships from macromolecular sequences. Today, in contrast, there are families of sequences for many genes. Journals such as *Journal of Molecular Evolution and Molecular Biology and Evolution* are largely devoted to the study of sequence evolution.

Recently, there has been an exciting new initiative in molecular biology, the Human Genome Initiative. This project has as its objective the characterization of an entire human genome of 3 × 109 bp. Even though the basic goal of the genome project is very important and will take several years to accomplish, the project has much broader implications and could be the basis of biology for the next century. First of all, the human genome cannot be considered in isolation from those of other organisms. For example, the genomes of *Escherichia coli*, nematode, and yeast are being characterized. With a fairly detailed restriction map of E. coli already in existence, both the experimental techniques and the basic biology learned from work on these organisms are essential to the Human Genome Initiative. From our point of view, genomic analysis is an important new emphasis in biology. Genomes from different organisms can be studied from a comparative viewpoint, and variation within a species is becoming accessible at the molecular level. When large amounts of sequence are produced, it will be an enormous task to detect functional and evolutionary relationships within and between the DNAs of organisms.

In this chapter we will consider sequence alignment, one of the most utilized computational tools for the study of molecular sequences. See Bell and Marr for related discussions of these topics. In the section on sequence alignment, we discuss the alignment problem, including a maximum-likelihood heuristic to motivate alignments that maximize a scoring function. The computer algorithms used to align full sequences and to locate maximum segment alignments are given. Alignments from 5S ribosomal RNA (rRNA) gene sequences are presented to illustrate the algorithms. The statistical significance of alignments is also discussed. Finally, there is a brief survey of multiple sequence alignments.

## Sequence Alignment

Several different questions might be asked about a sequence, one of which concerns unexpected relationships with other sequences. These discoveries are sometimes made by screening a database with the sequence. Wilbur and Lipman and Lipman and Pearson, devised a search algorithm based on the computer science technique of hashing, and theirs is the method of choice for such questions. Several groups have implemented versions of their technique. Essentially, the search looks for exact matching k-mers (usually $k = 4$ to 6 for DNA, and $k = 2$ for protein) and reports regions where the test sequence has a high density of matches with the database. Then an optimized alignment region is presented. There is no further discussion of database searches in this chapter, but they should not be overlooked.

Sequence alignment is a popular computer activity. The alignments are often based on some explicit optimization function, rewarding matches and penalizing mismatches, insertions, and deletions. Sequence alignment can give useful information about the evolutionary or functional relationships between sequences.

The generic two-sequence alignment problem can be stated as follows: a = $a_1 a_2 \ldots a_n$ and b = $b_1 b_2 \ldots$ bm are two sequences over a finite alphabet. For DNA and RNA, the alphabet has four letters, and for proteins, the alphabet has 20 letters. Molecular evolution events include substitutions of one letter for another as well as insertions and deletions (indels) of letters. What is the minimum number of these events required to transform a to b? The relationships between a and b implied by the minimal set of events are usually displayed as alignments where letters of a are written directly over their corresponding letters in b. It is necessary to weight the transformations in order to come closer to biology, so the problem has a more general statement.

Let us look at a simple example where a = AAGTIC and b = AGCCC.

An alignment that might be suggested by the sequences is

AAGTTC
A-GCCC

where there are three matches, two mismatches, and one indel. This alignment represents a specific hypothesis about the evolution of the sequences; three of the nucleotides have not changed since the common ancestor, there have been (at least) two substitutions, and one nucleotide has been either inserted or deleted. Sequences are aligned by computer to find good alignments. A computer is necessary because of the exponential number of possible alignments. For example, two sequences of length 1,000 have over $10^{600}$ possible alignments.

There remains the question of how to score an alignment. We present a simple heuristic derivation of "alignment score." If $p$ is the probability of an identity, $q$ the probability of a substitution, and $r$ the probability of an indel, the above alignment has probability

$$\Pr = p^3 q^2 r^1.$$

Define score $S'$ by the log likelihood:

$$S' \log \Pr = 3(\log p) + 2(\log q) + 1(\log r).$$

Define $S = S'n \log s = S' - 6 \log s$, where s is a constant satisfying log = 1. We have simply subtracted a constant from $S'$. $S$ becomes

$$S = 3 - 2\mu - 1\delta,$$

where $\mu$ = log and $\delta$ = log. Fortunately, score defined by

$S$ = #identities - $\mu$#substitutions - $\delta$#indels

can be efficiently computed as we will see in the next section. It also has the simple maximum likelihood interpretation we have presented. We caution the reader that this simple reasoning is only a heuristic. See Tavare for a related mathematically rigorous discussion.

We distinguish two types of alignments: (1) alignments of full sequences; and (2) finding segments of sequence that can be well aligned. Full sequence alignment should only be attempted when it is believed that the sequences are related, from one end to the other. If this is not the case, the sequences can be forced into incorrect relationships due to the necessity of matching the less similar segments. A more conservative approach is to run a maximum segments analysis that only finds those segments of the sequences matching at or above some preset level.

Consensus sequence analysis is usually done by the biologists "eye" and by experiment. Of course, we only believe a protein-binding site when it is verified by experiment, but analysis by inspection can be biased. So it is useful to have computer methods that can find consensus patterns best fitting explicitly stated criteria. Computer algorithms for consensus can be applied to align sets of multiple sequences.

## Dynamic Programming Alignment of Two Sequences

In 1970 Needleman and Wunsch wrote a landmark paper that approached sequence comparison (alignment) with an algorithm that is one of a class of optimization techniques known as dynamic programming. Their algorithm finds maximum similarity between two sequences, where matches receive positive weight, and mismatches, insertions and deletions receive non positive weight.

Some mathematicians began to attempt to define a distance between sequences and so to construct a metric space. This search culminated with Sellers, where the desired results were obtained for single insertions and deletions, and with later workers who extended this work to multiple insertions and deletions. While it has been proven that similarity and distance are equivalent concepts when matching full sequences for certain other problems similarity is superior, and we presented similarity here.

Modern DNA sequencing has revealed introns and exons. This suggests the problem of finding the best matching pieces (segments) between two sequences. While this problem was first approached with distance methods, a much simpler approach through similarity was taken by Smith and Waterman and an extension of that technique is presented below.

Matching or aligning entire sequences should be attempted when the sequences are known or suspected to be closely related. Even when this is the case, an extraordinary number of best alignments can result; many of these will differ only slightly from one another. We illustrate this below with some recommendations on how to deal with the situation. Most sequence comparisons will, however, involve sequences only significantly related in pieces, if at all. In those cases a full alignment is not informative, and the maximum segments algorithm is the most appropriate. These algorithms can produce segment matching which are best, second best, and so on; this is shown in the examples below.

## Aligning Full Sequences

As explained above, what is to be presented is a similarity method for sequence alignment. The function *s* is to define similarity between the letters *a* and *b*. In the examples below, matches receive weight 1 and mismatches receive - 1, so that $s(a, b) = 1$ if $a = b$ and $s(a, b) = -1$ if $a \neq b$. Deletions of length *k* receive weight- $w_k$ and below, only $w_1$ is used where $W_k = kw_1$. The algorithm is based on recursively computing a matrix *S*.

First

$$S_{0,0} = 0, S_{i,0} = - w_i, 1 = i = nS_{0,j} = - w_j, 1 = j = m.$$

For single insertions and deletions only,

$S_{i,j} = \max\{S_{i-1,j-i} + s(a_i, b_j), S_{i-1,j-1}\ w_1, S_{i,j-1} - w1\}$.

The idea of these calculations is that $S_{i,j}$ is the maximum similarity of $a_1 a_2 \ldots a_i$ and $b_1 b_2 \ldots b_j$. This is why for example, $S_{0,j} = w_j$. The recursion is based on the ways an alignment can end:

$\ldots a_i$ corresponds to $S_{i-1,j-1} + s(a_i, b_j)$

$\ldots b_j$

and

$\ldots -$ corresponds to $S_{i,j-1} - w_1$

$\ldots b_j$

and so on.

When n = *m*, the multiple insertions and deletions program runs in time proportional to $n^3$, while the single insertion and deletion program runs in the much preferred time $n^2$. Fortunately, when the function $W_k$ is linear in *k*, $w_k = \alpha + \beta * k$, the running time can still be made n2. Multiple insertions and deletions are important, as adjacent bases are deleted or inserted by one event and should be weighted accordingly.

Alignments can be produced from the matrix by two methods. One is accomplished by saving information at each matrix entry that indicates which events were necessary to calculate $S_{i,j}$. Then, beginning at $S_{n,m}$, this information is used to retrace our steps to $S_{0,0}$, thus producing the optimal alignments. The second technique is recomputing to see which events produced the value of each matrix entry beginning at $S_{n,m}$. Both methods take little time in comparison with the matrix construction.

For the examples, we first align genes of 5S rRNA *E. coli* (*rrnB* operon) with the closely related 5S rRNA *Beneckia harveyi*. The 5S rRNA gene sequences used in this chapter are taken from Erdmann. As mentioned above, $s(a, b) + 1$, if $a = b$, and $s(a, b) = -0.9$ if $a \neq b$ and $w_1 = 2$. ($w_k = \infty, k \geq 2$, so that only single insertions and deletions are allowed.) The two sequences have similarity of 78.1. There are two optimal alignments, which result from simply changing which of the two Ts is mismatched with the C.

For a second example, we align genes of 5S rRNA *E. coli* with the more distantly related 5S*Mycoplasma capricolum*. Here the similarity is 18.3, with 175 different but optimal alignments. The 175 alignments are represented in the four alignments of Figure 4-2.

The representation is similar, so that the first alignment actually represents a total of $2 \times 5 \times 3 = 30$ alignments, the second represents $2 \times 5 \times 2 = 20$, the third $5 \times 5 \times 3 = 75$, and the fourth $5 \times 5 \times 2 = 50$. This makes a total of $30 + 20 + 75 + 50 = 175$ alignments. These alignments have a great deal in common. As will be seen in the next section, a maximum segments algorithm will produce almost all of the common alignment.

## Maximum Segments

This is a preferred method for exploring sequence relationships if computer time is available. See Smith for database searches with the algorithm. It consists of a simple alteration of the preceding method due to Smith and Waterman. The recursively defined matrix here is $H$ and recursion begins with $H_{0,0} = H_{i,0} = 0, 1 = i = nH_{0,j} = 0, 1 = j = m$ and

For single insertions and deletions only,

$$Hi,j = \max \{H_{i-1,j-1} + s(a_i, b_j), H_{i-1,j} - w_1, H_{i,j-1} - w_1, 0\}.$$

Notice that a simple addition of zeros in the boundary conditions and recursion is the only change from the definition of $S$. These simple changes have the pleasant effect of causing Hi, j to be the maximum similarity of *all possible* segments ending in $a_i$ and $bj$. *Thus max* $_{ij} H_{ij}$ *is the score of the best aligned segments in a and b.*

There is no problem with the observation that the best segment matching is associated with the entries where max $_{ij}$ H $_{ij}$ is achieved. Since large entries influence the entries nearby, it is not clear whether or not the second-best matching is near the first or not. One way to make certain of finding the second best is to recalculate the matrix, only this time no match, mismatch, insertion, or deletion from the maximum segment can be used. With single or linear insertion and deletion functions, only a small part of the matrix need be recomputed.

## Statistical Distribution of Alignment Scores

Spurious matches are a problem in interpretation of the sequence alignments. In this section, we present some results for the distribution of scores maxH $_{ij}$ from the maximum segments algorithm.

The average sequence in GenBank or EMBL is 1,000 bases long. It shows best segment alignments for independent simulations of ten independent pairs of length 1,000 DNA sequences with $P(A) = P(C) = P(G) = P(T) = 0.25$. The algorithm parameters are $s = 1$ if $a = b$, $s = -\mu$ for $a \neq b$, and $w(k) = \delta k$, where $\mu = 1.5$ and $\delta = 2.1$. It is remarkable that these segmental matching from random sequences are so long and score so well. Simulations such as this suggest that understanding the distribution of score ($\max H_{ij}$) under the null hypothesis of independence is an important goal. If the analysis of "interesting" alignments proceeds on an ad *hoc* basis, it is easy to be misled by statistically insignificant alignments. As the genome projects get underway and megabases of sequence are produced, these statistical considerations will assume increasing importance.

## Multiple Sequence Alignment

Many problems of interest in biology involve more than two sequences. For example, the analysis of E. coli promoter sequences for regions of the sequences critical to polymerase binding to DNA involves all sequenced promoters. While analysis of a set of sequences for evolutionary relationships can be performed by pair wise sequence analysis, it is often essential to simultaneously consider the full set of sequences.

Alignment of $r > 2$ sequences by dynamic programming is possible in principle, but serious computational difficulties must be overcome. In Sankoff a technique is presented to optimally align r sequences given a tree that relates them. This method uses dynamic programming and Fitch parsimony algorithm. The method is very computationally intensive. Another method is that of Waterman in which a tree is not explicitly assumed. All sequences are treated symmetrically so that a so-called star phylogeny is implicitly assumed. These methods take at least time proportional to $n^r$ where n is the common sequence length.

Recently, some significant progress has been made in the area of multiple alignment by dynamic programming. Lipman and Carrillo have made it feasible to find optimal alignments of up to six sequences by greatly reducing the number of cells the dynamic programming algorithm must consider. Carrillo and Lipman give the basic algorithm. In Altschul and Lipman, it is extended to an algorithm that aligns the sequences given a tree that relates them. In Lipman up to eight sequences the length of an average protein can be aligned.

It is often important to analyze many sequences. A consensus algorithm for multiple DNA sequence alignment is given that matches words of length and degree of mismatch chosen by the user. As above, all sequences are treated symmetrically. The alignment maximizes an alignment scoring function. This method can easily align 100 sequences of length 1,000. In Waterman and Jones the method is extended to protein sequences.

A heuristic approach that has proven very useful is to infer the evolutionary tree while making a multiple alignment. Usually, the various methods proceed by performing pair wise alignments. In Johnson and Doolittle and Feng and Doolittle, one such development is pursued. Taylor has also written algorithms very useful for protein analysis. Many of these methods of multiple sequence analysis are presented in chapters of a comprehensive volume edited by Russell Doolittle. The program of Hein combines and refines many of the other techniques, and this would be our first choice to align and deduce an evolutionary tree for a set of sequences.

# 7

# DNA Databases for 21st Century

## Introduction

Mark Twain was speaking about fingerprints, of course, but prophetically, he might just as well have been speaking about DNA. The protagonist in Pudd'nhead Wilson was an attorney who had a passion for collecting fingerprints. One wonders if a modem-day version of Twain's vivid character would have the same affinity for collecting samples of DNA for law enforcement purposes. As law enforcement agencies the world over have been amassing huge collections of fingerprints since the closing days of the nineteenth century, so too have they recently begun to collect, organize, analyze, and store collections of DNA samples for forensic purposes.

This trend, as was the case with fingerprints, has been hailed as a godsend for crime fighting, but also decried as an evil at the same time. However, as with fingerprints, it looks like DNA testing and associated databases are here to stay. Accordingly, the current proliferation of DNA databases and their likely further expansion raise three significant policy issues and attendant questions. First, how do we utilize this new technology, while protecting against misuse and abuse? The question is really much more complex than this, and it certainly covers a multitude of sub-issues. At the essential core of this issue is the same question which appears in virtually every facet of our daily lives today. Science and technology are progressing at exponential rates, while the ordinary citizen struggles to keep up; so, what happens when technology, and the manifold advances it spawns, transcends society's ability to regulate such technology? Further, in the absence of a serious and well-informed debate about the advisability and demonstrative value of putting into practice whatever advances new technologies may provide, will particular interest groups exert unchallenged influence, if not complete hegemony in a particular area, and successfully lobby for very large expenditures of public financing?

In order to address the first issue properly, the investigation of a second issue requires careful and immediate attention. Although technology makes certain advances possible, are these advances truly necessary? Moreover, will they produce the alleged benefits and if so, at what cost? This paper will not attempt to solve this dilemma in a macro context. Rather, our interest centers on the forensic application of DNA technology, and in particular, the construction of scientific databases that contain such information. In the past few years, supporters of DNA testing for forensic applications have made remarkable claims about the potential of DNA testing as a crime fighting tool and have touted DNA as the next great breakthrough since fingerprints. It should be noted at the outset that we have no quarrel with DNA testing per se. The scientific community has conclusively demonstrated the reliability and validity of DNA testing and that the "matching" of an evidence sample with that taken from a suspect for purposes of exclusion versus inclusion can be highly successful. Further, although at one time there was considerable debate about the admissibility of DNA evidence, the point is now moot.

However, our inquiry is guided by a healthy skepticism about the widespread collection of DNA samples and their subsequent storage in databases as a crime control measure. For example, a new program in New York, announced as recently as August 7, 1999, requires (as of December 1, 1999) that any person convicted of certain designated felony offenses will have his/her DNA stored in a database. The program will involve both persons newly convicted (about 25,000 persons per year) and the retroactive testing of persons already in the criminal justice system (approximately 100,000 prison inmates). Estimates of the cost of the program range from $10 million to $20 million.

Regarding the potential of such programs as a law enforcement tool, Ronald S. Neubauer, the president of the International Association of Chiefs of Police offered the following comments:

> *I think it's one of the most important developments in forensic science in law enforcement.... And in the 21st century, I not only see DNA being a tool to solve crimes, but as a way to insure that innocent people are not being convicted of crimes they did not commit.*

It is precisely this type of unsupported assertion, if not blatant exaggeration, concerning the crime fighting value of DNA, together with the manifold costs of DNA testing and database construction, that frame the scope of our analysis. To put it more simply, will DNA databases provide law enforcement and the subsequent criminal prosecutions with measurable and significant effects on crime? Further, can these effects, once demonstrated and replicated on a wide-scale basis, be produced in a cost-effective manner? Assuming that DNA databases are indeed valuable in the fight against crime, and can be administered in a cost-effective fashion, a final remaining issue arises concerning appropriate regulations surrounding DNA database construction, maintenance, and access. As will be shown below, since there are various schemes concerning who should be required by statute to contribute DNA samples, this question indeed poses significant legal and ethical issues which must not be ignored or dismissed amidst the fervor surrounding the alleged benefits of DNA testing.

## Emergence of DNA Typing

**A. DNA is Disovered.** The DNA story begins with two gentlemen named Watson and Crick who came upon a remarkable discovery in 1953. They unraveled the mystery of DNA for the first time, obtaining a Nobel Prize for their efforts. Like many scientific discoveries, it would take years to realize the full magnitude and potential of this pioneering work. It was not until the early 1980s that Dr. Alec Jeffreys at the University of Leicester in England pioneered the use of DNA in the law enforcement arena. The Federal Bureau of Investigation quickly followed suit in 1988.

DNA is the chemical deoxyribonucleic acid, which carries the genetic code of each human's body – the genetic blueprint we inherit from our parents. DNA, while not actually a part of saliva, urine, perspiration, or tears, is found in one place, and only one place—the nucleus of cells. Because these cells are found in all bodily fluids, tissue, and hair, DNA is an omnipresent residue that trails us wherever we go. These physical properties of DNA have made it an important tool in fighting crime. Presently, there are three principal methods by which DNA testing is usually accomplished: (1) Restriction Fragment Length Polymorphism (RFLP); (2) Polymerase Chain Reaction ; and (3) Short Tandem Repeats (STRs). Depending upon the quantity and quality (i.e., molecular weight and possible degradation) of the forensic sample available, the time frame available for testing, and other factors, one or more of these methods will generally produce valid results for making a "match" between an evidence sample and a suspect sample for purposes of excluding or failing to exclude the suspect as the perpetrator.

**B. Federal DNA Initiatives.** As might be expected, the United States Department of Justice has taken the lead in developing DNA applications for use by law enforcement agencies and subsequent criminal case disposition. The DOJ has utilized both the FBI and the research and funding capabilities of the National Institute of Justice to develop a far reaching set of programs for DNA applications. The FBI, as noted above, began its involvement with DNA in 1988. Since then, the FBI has continued to play a leading role and has developed a number of DNA initiatives.

First, the FBI maintains two DNA analysis units. DNA Analysis Unit I performs laboratory testing on evidence samples taken from violent crime scenes. Body fluids and fluid stains are examined serologically and then the DNA is characterized through RFLP and/or PCR testing. In 1996, the FBI opened another unit, DNA Analysis Unit II, which uses mitochondrial DNA testing on evidence samples when the sample is degraded or contains an insufficient amount of DNA for either RFLP or PCR testing. Second, the FBI operates the Combined DNA Index System (CODIS) which began as a pilot project in 1990. CODIS is a software-based system which uses two indices to facilitate violent crime investigations. Third, the DNA Identification Act of 1994 authorized the FBI to establish further DNA indices. The Convicted Offender Index contains DNA profiles of felons convicted of violent crimes and sex offenses, while the Forensic Index contains DNA profiles from crime scenes. The CODIS software permits DNA samples across the two indices to be compared for possible matches thereby facilitating criminal investigations. Fourth, in October 1998, the FBI announced its newest tool – the National DNA Index System (NDIS). The NDIS is an electronic system that will allow federal, state, and local law enforcement agencies to contribute DNA samples to the national database and thereby enhance the investigation of violent crimes. Fifth, the FBI will soon inaugurate a Federal Convicted Offender DNA Database. The Anti-Terrorism and Effective Death Penalty Act of 1996 authorized the FBI to implement a supplement to CODIS by requiring federal prisoners, who are convicted of an offense against a minor or a sexually violent offense to provide a DNA sample prior to release from a federal correctional institution.

The Department of Justice has also been providing significant leadership on DNA applications for law enforcement through the National Institute of Justice . NIJ provides research and program funding on a wide range of topics, including DNA, through its Science and Technology division. Since 1986, NIJ has produced a number of significant publications on DNA and has funded numerous projects on improving DNA technology. Arguably, an NIJ grant to the Institute for Law and Justice achieved a milestone in DNA applications for forensic purposes. This pioneering study investigated the use of DNA testing to exonerate convicted offenders through a process of post-conviction relief. Attorney General Janet Reno was so impressed by the results of the study that she directed NIJ to establish a National Commission on the Future of DNA Evidence.

The Commission's agenda addresses five areas with respect to DNA: (1) DNA in post-conviction; (2) legal issues on admissibility and discovery; (3) training for crime scene collection; (4) laboratory capabilities; and (5) the impact of future technologies. The Commission has held five meetings since its inception in March 1998. The work of the Commission is available in the Proceedings of the five meetings, which span some 500 pages.

## DNA Effectiveness: Theory vs. Reality

**A. How DNA Databases are Theoretically Supposed to Work.** In theory, a DNA database consists of DNA samples obtained from two sources: crime scene evidence and individual "donors." The term "donors" is used simply for utility here, and as Part V.B will show, who is included in the donor group varies substantially from jurisdiction to jurisdiction. It should be noted here that other sources of DNA might provide samples for these databases, such as unidentified human remains or the DNA of relatives of missing persons. However, these records typically constitute only a small part of the average DNA database, and consequently, will be disregarded for the purposes of this paper.

Once a DNA sample intended for storage in a DNA database is obtained, it is sent to a DNA laboratory for processing. Once it has been analyzed, the results are stored in a central database. After this process is completed, the results of every DNA specimen in that database can be compared with every other sample in the database, and these samples can also be checked against new samples taken from people, crime scenes, or otherwise obtained elsewhere.

The ultimate value of a crime-fighting measure depends, not upon theory, or exaggerated speculations, or even anecdotal accounts, but rather on the real-world effectiveness of the technique. Thus, collection and study of empirical data is crucial to evaluating the advantages and disadvantages of such methods. The remainder of this section will first analyze the effectiveness of DNA databases from a crime-level or "pure effectiveness" standard and then consider the additional factor of cost to determine the ultimate value of DNA databases.

**B. Pure Effectiveness.** Clearly, because the DNA databases are relative newcomers to the fight against crime, they have yet to make a significant impact on crime rates. This should not, however, be viewed as a failure, at least at this early juncture in their history. Due to the very nature of a database, its utility theoretically increases proportionally as the amount of data contained in it expands. Therein lies the problem: DNA analysis is still quite time consuming, and this has led to a massive backlog of unanalyzed samples in our nation's crime laboratories. This backup may contain as many as 450,000 samples waiting to be analyzed. As of the date of this writing, the Texas Department of Public Safety had analyzed just over one-haft of the samples it had received. As these backlogs begin to recede, the effectiveness of the databases will continue to rise. Crime labs across the country are continuing to expand and upgrade their existing technology, and the technology itself is rapidly progressing.

Despite the backlogs, anecdotal evidence has already demonstrated that DNA databases can have remarkable effectiveness in solving crimes. For example, while these databases were still in their infancy, a brutal rape-murder in Illinois was solved using the Illinois DNA database. A young man was murdered, and his wife was raped, shot point-blank in the head, and left for dead. More than a year and a half later, the police had no solid leads. Forensic specialists obtained a DNA profile from the perpetrator's semen, and compared that with the Illinois DNA database. At the time, this database had only 600 samples on file from convicted sex offenders. Despite this handicap, the crime scene evidence matched DNA from a man who had already served time for sexual assault.

In Texas, the first "cold hit" occurred in May 1998, when DNA evidence recovered from crime victims was matched against a DNA record on file in the Texas DNA database. In this case, two young Granbury girls were sexually assaulted in 1993 in a dry creek bed near their home. Authorities had no leads on the case. However, the perpetrator was subsequently convicted of a 1995 sexual assault. As a result of the conviction, his DNA was sampled and included in the Texas DNA database. The crime scene evidence from the 1993 incident was run against the state database and it matched the suspect. After further tests were taken to confirm the match, the suspect, who was still serving time for the 1995 assault, confessed to the rapes.

Nationwide, the FBI database has produced some 583 cold hits to date. The National FBI database (CODIS) currently contains DNA samples from over forty-two states, and is expected to include DNA from all fifty states within the near future. England, which started its DNA database before the FBI, has had over 28,000 matches. The scope of Britain's database is considerably wider in terms of whose DNA is included than that of most United States jurisdictions, and it already has more than 360,000 samples indexed and on file. It is expected to eventually encompass more than one third of all English men between the ages of sixteen and thirty.

These isolated successes of DNA databases are interesting and laudatory. However, they do not provide systematic, conclusive, and widespread evidence that such databases, especially the expanded or "all-inclusive" variety, will be proven useful in the fight against crime. Are we to conclude that these few examples will increase dramatically as the databases proliferate and become interconnected? In order to address this fundamental question, we provide data below that exposes the usefulness of DNA databases to a proper scientific test on data concerning law enforcement as well as prosecutorial functions in the fight against crime.

## DNA Effectiveness: Law Enforcement

There are two basic applications for DNA in law enforcement, and these two widely divergent applications must be differentiated so that the proper focus of our inquiry will be clear. First, there is DNA testing concerning known suspects and evidence samples. Here, the DNA extracted from bodily fluids or tissue found at a crime scene (e.g., blood or semen), or a victim's DNA extracted from residue left on an offender (e.g., the victim's blood) are compared to determine if there is a match. It would seem that in the absence of other explanatory information, a DNA match or non-match would be dispositive of the suspect's involvement in or his/her innocence of the crime. We wholeheartedly and unequivocally endorse this particular use of DNA testing with known offenders, and further, encourage its use as broadly as possible. The only meaningful caveats we would offer involve proper training for crime-scene technicians and laboratory personnel as well as sound certification policies and well-conceived oversight and monitoring processes for both evidence collection and subsequent DNA testing.

However, a second (and highly touted) use of DNA concerns the construction of massive DNA databases to facilitate what we shall refer to as the "DNA mining process." As noted above, the logic behind DNA databases appears convincing, and concomitantly, such databases are touted as major crime fighting tools. It would seem to make sense that all that society needs do to fight crime effectively is: (1) capture the DNA from known offenders (the exact selection of offenders remains open to debate); (2) store the DNA in a database; and (3) compare the offender bank DNA with that taken from crime scenes. The promised results of course will be the identification and subsequent arrest of a suspect and his or her successful prosecution owing to the DNA match, a result which would not have been possible but for the DNA database. However, when one examines the nature and distribution of crime, the presumed usefulness of DNA databases as a crime control measure may not only be far from obvious or certain, but may turn out to be grossly exaggerated. We thus turn to a consideration of crime events and their susceptibility to DNA applications in law enforcement. In 1997, there were 13,175,100 Index offenses reported to the FBI. Index offenses are deemed to be the most serious offenses and are classified as Index crimes, or Part I crimes. Violent Index crimes, which number 1,634,770 offenses, account for only 12.4% of all Index crimes. The vast majority of Index crimes, some 11,540,300 offenses, are crimes against property, accounting for 87.6% of all Index offenses. Further, the rates (per 100,000 persons) clearly indicate that people are at much greater risk of being victims of property offenses than violent crimes, considering that property crimes occur at a rate of 4,311.9 per 100,000, a rate which is seven times higher than that for violent Index crimes (610.8 per 100,000).

**Table 7.1. Crime Index Offenses Reported to FBI in 1997.**

| Index | Crimes Offense | Percent | Rate per Reported of Total 100,000 |
|---|---|---|---|
| Murder/Non-Negligent Manslaughter | 18,210 | 0.1% | 6.8 |
| Forcible Rape | 96,120 | 0.7% | 35.9 |
| Robbery | 497,950 | 58.6% | 186.1 |
| Aggravated Assault | 1,022,490 | 7.8% | 382.0 |
| Violence Subtotal | 1,634,770 | 12.4% | 610.8 |
| Burglary | 2,461,100 | 18.7% | 919.6 |
| Larceny | 7,725,500 | 56.8% | 2,886.5 |
| Motor Vehicle Theft | 1,353,700 | 10.3% | 505.8 |
| Property Subtotal | 11,540,300 | 87.6% | 4,311.9 |
| Total Index Crimes | 13,175,100 | 100.0% | 4,922.7 |

The vast majority of serious crime is committed against property and not people. Accordingly, the vast majority of law enforcement responses consist of the following: responding to the scene, conducting interviews for purposes of writing a contact report, and subsequent follow-up investigation (usually by detectives). Unfortunately, the overwhelming majority of all police responses involve a property offense in which the victim seldom, if ever, sees or confronts the offender, and very likely has no idea who the offender may be. Further, it is highly likely that these voluminous property offenses do not occur with much trace evidence (even fingerprints) left at the scene, evidence which the police actually collect and upon which they can subsequently base their investigation.

The lack of physical evidence in property crimes further suggests that Index crimes will likely have differential arrest rates owing to the circumstances surrounding the event, particularly the availability of witnesses or the face-to-face victimization of violent offenses as opposed to the impersonal victimization of property offenses. The data in Table 7.2 represent the Index offenses "cleared" by the arrest of a suspect and attest to the general ineffectiveness with which law enforcement solves major crimes. As would be expected, the police are more successful in arresting suspects in violent crimes (average of 48.3%; 66.1% of murder and non-negligent manslaughter, 50.8% of rapes, 26.3% of robberies, and 58.5% of aggravated assaults) than in property crimes (average 17.5%; 13.8% of burglaries, 19.8% of larceny/theft, and 14.0% of vehicle thefts).

**Table 7.2. Crime Index Offenses Reported to FBI and Cleared by arrest in 1997.**

| Index Offense | Crimes Known | Percent Cleared byArrest |
|---|---|---|
| Murder/Non-Negligent Manslaughter | 14,759 | 66.1% |
| Forcible Rape | 78,975 | 50.8% |
| Robbery | 411,137 | 26.3% |
| Aggravated Assault | 838,711 | 58.5% |
| Violence Subtotal | 1,343,642 | 48.3% |
| Burglary | 2,044,918 | 13.8% |
| Larceny | 6,392,542 | 19.8% |
| Motor Vehicle Theft | 1,147,391 | 14.0% |
| Property Subtotal | 9,584,841 | 17.9% |
| Arson | 76,018 | 17.5% |
| Total Index Crimes | 11,004,501 | 21.6% |

III, AT 214 The property offense data are especially noteworthy. The data indicate that 9,584,841 property Index crimes were reported by police agencies that also reported clearance data. Despite this substantial crime volume, only 17.9% were solved by arrest; therefore, 7,869,154 offenses did not result in the arrest of a suspect.

We come then to the issue of Effectiveness Test #1. In order for the UCR Index crimes to represent viable candidates for being solved by "DNA mining," there would have to be careful and painstaking crime scene investigation. In particular, the crime scene response would have to include forensic technicians and crime scene technicians (or "criminalists," as they are often called) who would scour the crime scene looking for trace evidence like blood, other bodily fluids, tissue, hair, etc., which carries the DNA of the perpetrator. Thus, the success of the DNA mining expedition for crime-fighting depends on three fundamental prerequisites: First, the criminal has to leave evidence behind at the crime scene, or on the person or clothing of the victim, that contains the criminal's DNA. Second, a trained technician must search the crime scene for this evidence. Third, the DNA-bearing evidence has to be, in fact, found, collected, and be of sufficient quantity and quality to permit DNA testing.

Let us be realistic here about the likelihood of these three prerequisites actually taking place, rather than permit ourselves to be swept up in the euphoria exemplified by the proponents of DNA mining. What do the groups of serious offenses classified by FBI as Index offenses tell us? The answer is straightforward – that DNA databases will not be greatly successful in increasing the extent to which police solve the vast majority of Index crimes. There are two principal and inescapable reasons for this conclusion. First, law enforcement already does a more than creditable job (i.e., greater than 50% clearance) of solving three out of the four violent Index crimes (66.1% of murders; 50.8% of rapes; and 58.5% of aggravated assaults). Second, as we have shown, the vast majority of Index crimes are property offenses, and this offense type does not carry a high potential for beneficial DNA testing, owing to the fact that the usual property offense crime scene is not likely to have the perpetrator's DNA, and even if it does, such evidence will hardly be looked for, let alone collected and tested for comparison to the databases.

b. Effectiveness Test #2: Non-Index crimes

The argument could be made that Index crimes are not the only offenses worthy of consideration, that there are countless other felonies which come to the attention of the police, for which DNA mining would be beneficial. These other offenses are known as "Non-Index" crimes. They are deemed to be less serious than Index offenses, and, accordingly, the FBI does not publish counts concerning the number of such offenses that are reported to the thousands of law enforcement agencies across the country. The FBI does, however, publish data concerning the arrests of suspects for Non-Index events. Table 7.3 provides FBI estimates of the number of persons arrested in 1997 for the Index crimes and the large category referred to as Non-Index crimes.

**Table 7.3. Estimated Arrests for Index and Non-index Offensesreported to FBI in 1997.**

| Index Offense | Persons Arrested | Percent of Total |
|---|---|---|
| Violence | 717,750 | 4.7% |
| Property | 2,015,600 | 13.0% |
| Arson | 20,000 | 0.1% |
| Total Index Arrests | 2,733,400 | 17.8% |
| Non-Index Offense | | |
| Against Person | 1,653,600 | 10.7% |
| Other | 11,004,100 | 71.5% |
| Total Non-Index Arrests | 12,547,700 | 82.2% |
| Total Index Crimes | 15,284,300 | 160.0% |

The arrest data indicate that only 17.8% of all arrests in the United States involve Index crimes, and further, only 4.7% of total arrests concern violent Index crimes, while 13.0% involve property Index offenses. The vast majority of arrests, 82.2%, concern Non-Index, or less serious, offenses; of these, only 10.7% concern crimes against persons. Thus, even if we ignore the Index vs. Non-Index distinction, only 15.4% of all arrests concern a violent crime (however serious or minor) against a person.

Table 7.4 provides detail about the distribution of arrests across the Index crimes. Clearly, law enforcement activity is substantially devoted to the arrest of persons for crimes other than those involving violence against one or more victims. These data show that only 26.3% of arrests for Index crimes involve violence against a personal victim. Similarly, Table 5 provides identical data on the arrests made for Non-Index offenses. Again, arrests for violent Non-Index offenses account for only 13% of all the arrests for Non-Index crimes. The remaining 87% of Non-Index arrests concern a variety of criminal behaviors that are not serious and do not involve any type of assaultive or violent behavior against a person.

**Table 7.4. Estimated Arrests for Crime Index Offenses Reported to FBI in 1997.**

| Index Offense | Persons Arrested | Percent of Total |
|---|---|---|
| Murder/Non-Negligent Manslaughter | 18,290 | 0.6% |
| Forcible Rape | 32,060 | 1.2% |
| Robbery | 132,450 | 4.8% |
| Aggravated Assault | 534,920 | 19.6% |
| Violence Subtotal | 717,750 | 26.2% |
| Burglary | 356,000 | 13.0% |
| Larceny | 1,472,600 | 53.9% |
| Motor Vehicle Theft 167,000 | 6.1% | |
| Property Subtotal 2,015,600 | 73.0% | |
| Arson | 20,000 | 0.7% |
| Total Index Crimes | 2,733,320 | 100.0% |

Regarding Effectiveness Test #2, the conclusion is as inescapable as was the case for Test #1. Law enforcement activity produces arrests for Non-Index crimes, which involve "other" crimes much more often than for violent crimes . To what extent, then, will DNA mining be beneficial for the vast majority of Non-Index crimes? Realistically, the answer must be that DNA testing is quite irrelevant for these offenses. These are, by all accounts, less serious crimes, and local law enforcement hardly has the necessary resources to treat these offenses as though they require the intensive crime scene effort that is usually reserved for violent crimes against the person.

**Table 7.5. Estimated Arrests for Non-index Offenses Reported to FBI in 1997**

| Non-Index Offense | Persons Arrested | Percent of Total |
|---|---|---|
| Against Person | | |
| Other Assaults | 1,395,800 | 11.0% |
| Sex Offenses (not Rape) | 101,900 | 0.8% |
| Offenses Against Family & Children | 155,800 | 1.2% |
| Other | | |
| Curfew | 182,700 | 1.4% |
| Disorderly Conduct | 811,100 | 6.4% |
| Driving Under Influence | 1,477,300 | 11.7% |
| Drugs | 1,683,600 | 13.3% |
| Drunkenness | 734,800 | 5.8% |
| Embezzlement | 17,400 | 0.1% |
| Forgery & Counterfeiting | 120,100 | 0.9% |
| Fraud | 414,600 | 3.3% |
| Gambling | 15,900 | 0.1% |
| Liquor Laws | 636,400 | 5.0% |
| Prostitution | 101,600 | 0.8% |
| Runaways | 196,100 | 1.5% |
| Suspicion | 6,500 | 0.1% |
| Stolen Property | 155,300 | 1.2% |
| Vagrancy | 28,800 | 0.2% |
| Vandalism | 318,400 | 2.5% |
| Weapons | 218,900 | 1.7% |
| All Other Offenses | 3,884,600 | 30.7% |

An alternative perspective on the availability of DNA-bearing evidence and its collection potential, however, has been advanced elsewhere. Weedn and Hicks have noted that "at most crime scenes, there are many kinds of biological evidence: not only blood and hair but also botanical, zoological, and other types of substances." The authors advance their argument by citing data collected in one study revealing that blood evidence was found in 60% of murders and in a similar percentage of assaults and batteries, while hair was found at the scene of 10% of robberies and 6% of residential burglaries. It is obvious, of course, that the authors' use of the word "most" above is sheer hyperbole. There is no conceivable way that DNA-related evidence found at the scene of only 10% of robberies and 6% of burglaries can be considered to qualify as "most" crimes. In fact, the data would suggest the alternative – that DNA-bearing evidence is available at only a small number of crime scenes. It is likely, of course, that the authors are really arguing that DNA related material is theoretically available at some crime scenes and could be and should be collected more often than is presently done, thereby aiding the investigation. Thus, they note that:

> *saliva, skin cells, bone, teeth, tissue, urine, feces, and a host of other biological specimens, all of which may be found at crime scenes are also sources of DNA. Saliva may be found in chewing gum and on cigarette butts, envelopes, and possibly drinking cups. Fingernail scrapings from an assault victim or a broken fingernail left at the scene by the perpetrator may also useful DNA evidentiary specimens. Even hatbands and other articles of clothing may yield DNA.*

We readily accept the notion that there may be biological specimens left at some crime scenes that could quite readily yield DNA for testing. But, we repeat our admonition outlined above: while there may be such evidence available at crime scenes, it is highly unlikely that police departments have sufficient resources to look for such evidence across the wealth of crimes with which they must concern themselves. Weedn and Hicks are well aware of this as they state "little of this evidence is recovered from crime scenes, less is submitted to crime labs, and still less is analyzed." The reasons are obvious. It is often difficult enough to convince the police to dust for fingerprints at a residential burglary because the police know that their search will likely be futile. Imagine, therefore, trying to convince police to search the crime scene (usually outside) of a robbery for such evidence as the perpetrator's hair, tissue, or other biological residue. Or imagine further, a crowded parking lot outside a bar or night club that was the scene of an assault. Here, the police would be expected to hunt for a broken fingernail or the victim would have to turn over his or her clothing so that a search for trace evidence could be conducted. Thus, it is unrealistic to argue that DNA is readily available at crime scenes if only the police would look for it. Assuming that the legitimate law enforcement reluctance to treat every crime scene, regardless of how serious, as if it were a homicide or a rape could be overcome, where will the resources come from to look for the biological evidence, collect it, test it, and store it in a DNA database? It must be recognized that the front end of the process, the crime scene, is highly labor intensive, and it is doubtful that local law enforcement has such resources available. Weedn and Hicks have specifically noted that the scarce resources available to law enforcement which must be distributed across a range of pressing needs is a significant limitation to DNA as a crime-fighting tool. Moreover, not only is the process highly labor intensive, that labor tends to be highly trained, and therefore, more expensive. Further, even if such resources could be made available somehow, the question arises as to whether this is a wise expenditure of public resources. It should be stressed that we do not wish to suggest abandoning current forensic evidence collection practices, but only that there are significant and numerous obstacles to realizing substantial crime-reduction results from this process.

## DNA Effectiveness: Prosecution

The other stage of the criminal justice system for which DNA mining (and the DNA databases) could be most beneficial concerns the prosecution of criminal defendants. Just because DNA mining might not dramatically increase the rate at which crimes are solved by the police, there are still advantages to the prosecutorial use of DNA results. DNA could produce the following benefits: (1) convictions could be more likely; (2) convictions without undue plea-bargaining could occur more often; (3) some, if not many, defendants may even plead guilty in the face of such strong forensic evidence; (4) some defendants who might not otherwise be brought to trial could be convicted; and (5) some suspects would be exonerated before trial, thus sparing the necessity of a trial – both the expense as well as unnecessary discomfort and embarrassment for the wrongfully accused. Naturally, this list is not meant to be exhaustive, but merely suggestive of DNA's potential in court. It would seem, therefore, that DNA use by prosecutors would be universally high as it would be difficult to assemble reasons to the contrary.

Table 7.6 provides the results of three years worth of data from an ongoing Bureau of Justice Statistics study of the nation's prosecutors' offices. The data displayed are from the last three installments of the research: 1992, 1994, and 1996. Overall, the results indicate that the use of DNA by prosecutors increased from 25% in 1992 to 42% in 1994, then to 49.2% in 1996. In the 1994 survey, the data were available by size of office, and these results show that when the prosecutor's office serves a very large population (500,000 citizens or more), almost all such offices use DNA as compared to the offices serving smaller populations (47% for populations under 500,000, and 23% for a part-time office). The 1996 survey provides a different population breakdown with four categories rather than three. Again we see that as the population served increases, there is a corresponding increase in DNA use by the prosecution. The use of DNA ranges from 100% for populations of one million or greater, to 97.6% for populations of 250,000-999,999, to 56.4% for jurisdictions with populations under 250,000, and finally, to 15.3% for part-time offices.

**Table 7.6. Prosecutorial Use of DNA Evidence: 1992-1996**

| | Percent of Offices Using DNA | | |
|---|---|---|---|
| **Population Served** | **1992** | **1994** | **1996** |
| All Offices | 25% | 42% | 49.2% |
| 500,000 or More | n.a. | 95% | n.a. |
| Under 500,000 | n.a. | 47% | n.a. |
| Part-Time Office | n.a. | 23% | n.a. |
| 1,000,000 or More | n.a. | n.a. | 100.0% |
| 250,000 to 999,999 | n.a. | n.a. | 97.6% |
| Under 250,000 | n.a. | n.a. | 56.4% |
| Part-Time Office | n.a. | n.a. | 15.3% |

John M. Dawson et al., Prosecutors in State Courts, 1992, BUREAU JUST. STAT. BULL., Dec. 1993, at 1; Carol J. DeFrances et al., Prosecutors in State Courts, 1994, BUREAU JUST. STAT. BULL., Oct. 1996, at 4; Carol J. DeFrances et al., Prosecutors in State Courts, 1996, BUREAU JUST. STAT. BULL., July 1998, at 6. Table 7.7 provides data from the 1996 survey concerning how prosecutors used DNA evidence. Overall, the results show that less than half of all prosecutors (49.2%) use DNA in any manner. More particularly, only 41.1% use DNA evidence during plea negotiations, while even less, 34.2%, use DNA at trial. With respect to DNA use by type of crime charged, the data once again suggest that use is limited and generally restricted to two offenses. That is, 27.5% of prosecutors use DNA evidence for murder and non-negligent manslaughter cases, while 42.7% use such evidence in sex offenses. The remaining three offense types show extremely low levels of DNA use: aggravated assault – 4.2%; robbery – 2.9%, and burglaxy – 2.7%.

**Table 7.7. Prosecutorial Use of DNA Evidence By Population Group, Stage of Case, ond Offense, 1996**

| | Population Served by Prosecutor | | |
|---|---|---|---|
| **Use of DNA** | **All** | **1,000,000+** | **250,000 to 999,999** |
| Used Any Time | 49.2% | 100.0% | 97.6% |
| Plea Negotiations | 41.1% | 90.9% | 84.1% |
| Trial | 34.2% | 100.0% | 91.8% |
| Murder/Manslaughter | 27.5% | 93.8% | 87.5% |
| Sex Offense | 42.7% | 100.0% | 96.4% |
| Aggravated Assault | 4.2% | 36.4% | 13.7% |
| Robbery | 2.9% | 15.2% | 13.7% |
| Burglary | 2.7% | 21.2% | 12.6% |

| | Population Served by Prosecutor | |
|---|---|---|
| **Use of DNA** | **Less Than 250,000** | **Part Time** |
| Used Any Time | 56.4% | 15.3% |
| Plea Negotiations | 48.3% | 9.8% |
| Trial | 36.7% | 9.8% |
| Murder/Manslaughter | 28.8% | 4.2% |
| Sex Offense | 46.5% | 15.3% |
| Aggravated Assault | 4.2% | 0.0% |
| Robbery | 2.6% | 0.0% |
| Burglary | 2.3% | 0.0% |

**Carol J. DeFrancis & Greg W. Steadman, Prosecutors in State Courts, 1996, BUREAU JUST. STAT. BULL., July 1998, at 6. There are, however, two basic findings that are readily apparent in the data when population groupings are considered. These findings suggest population differentials in the use of DNA evidence, and**

they also suggest that there are distinct crime-type differences. First, like the result concerning whether DNA was used at all, the type of use varies by size of population served. The two larger population groups (1 million or more and 250,000-999,999) make extensive use of DNA at both trial and during plea negotiations. On the other hand, the two smaller groupings (less than 250,000 and part-time offices) are much less likely to use DNA at trial or during plea negotiations. Second, the data also display very significant crime type variation in the use of DNA. That is, the two larger population groups use DNA evidence at a substantial rate for murder or non-negligent manslaughter (93.8% and 87.5%) and sex offenses (100% and 96.4%), but the two smaller population groups do not use DNA evidence in the majority of cases for either murder or non-negligent manslaughter (28.8% and 4.2%) or sex offenses (46.5% and 15.3%). Additionally, the data confirm our suspicion that other crime types beyond homicides and rapes are not readily susceptible to DNA evidence. That is, for neither aggravated assault, robbery, nor burglary do any of the prosecution groups exceed 36.4% in the use of DNA. In fact, from 79% to 100% of the time, the various offices do not use DNA evidence for two major violent crimes (robbery and aggravated assaults) or a major property crime (burglary).

These results are highly pertinent to the question of DNA effectiveness, as even those prosecution offices (i.e., the two largest groups) that use DNA evidence extensively do not use such evidence beyond the two most serious crime types. One can only wonder why this is the result. Is it because such evidence is not collected or tested and, therefore, is not available? Or, could it be that the crime types in question do not require DNA evidence to make a case sound enough to go to trial?

It might be argued that the variation in DNA use across population groups is not very important because Table 7.7 indicates that at least the largest counties use DNA evidence from about 84% to 100% of the time. Moreover, according to the Bureau of Justice Statistics, the larger cities and counties (those with populations greater than 500,000) represent about 49% of the total United States population. By implication, therefore, at least one-half of crimes are being prosecuted by offices that use DNA evidence extensively. Since BJS does not provide a list of these 127 prosecutors' offices, we cannot perform a direct test of the population covered. However, we can use FBI data to perform an approximate test, giving every benefit of the doubt to the BJS claims. These data are displayed in Table 7.8.

**Table 7.8. Offenses Reported to the FBI by Population Group and Index Offense Type**

| Population Group | Total Index | Violent | Property |
|---|---|---|---|
| All 220,142,000 | 11,097,248 | 9,689,324 | 9,689,324 |
| Group I | | | |
| 1 mil. | 1,422,710 | 294,349 | 1,128,361 |
| 20,922,000 | 12.8% | 20.9% | 11.6% |
| 500-999K | 872,274 | 130,140 | 742,134 |
| 11,084,000 | 7.9% | 9.2% | 7.7% |
| 250-499K | 1,209,667 | 188,962 | 1,020,705 |
| 13,825,000 | 10.9% | 13.4% | 10.5% |
| Group H 100-249K | 1,530,238 | 187,752 | 1,342,486 |
| 22,084,000 | 13.8% | 13.3% | 13.9% |
| Group HI 50-99K | 1,169,925 | 131,733 | 1,038,192 |
| 21,279,000 | 10.5% | 9.4% | 10.7% |
| Group IV 25-49K | 1,009,981 | 96,307 | 913,674 |
| 20,878,000 | 9.1% | 6.8% | 9.4% |
| Group V 10-24K | 968,773 | 81,972 | 886,801 |
| 22,577,000 | 8.7% | 5.8% | 9.2% |
| Group VI > 10K | 806,401 | 64,668 | 741,733 |
| 18,279,000 | 7.3% | 4.6% | 7.7% |
| Suburban Counties | 1,599,605 | 176,619 | 1,422,986 |
| 46,310,000 | 14.4% | 12.5% | 14.7% |
| Rural Counties | 507,674 | 55,422 | 452,252 |
| 22,903,000 | 4.6% | 3.9% | 4.7% |
| **Population Group Murder/ Rape Aggravated** | | | |
| **Manslaughter Assault** | | | |
| All 220,142,000 | 15,685 | 78,196 | 871,135 |
| Group I | | | |
| 1 mil. | + 3,372 | 7,502 | 162,268 |
| 20,922,000 | 21.5% | 9.6% | 18.6% |
| 500-999K | 1,685 | 6,929 | 70,587 |
| 11,084,000 | 10.7% | 8.9% | 8.1% |

| | | | |
|---|---|---|---|
| 250-499K | 2,240 | 9,227 | 108,059 |
| 13,825,000 | 14.3% | 11.8% | 12.4% |
| Group H 100-249K | 2,164 | 10,172 | 111,067 |
| 22,084,000 | 13.8% | 13.0% | 12.7% |
| Group HI 50-99K | 1,184 | 8,232 | 81,647 |
| 21,279,000 | 7.5% | 10.5% | 9.4% |
| Group IV 25-49K | 785 | 6,787 | 63,368 |
| 20,878,000 | 5.0% | 8.7% | 7.3% |
| Group V 10-24K | 684 | 6,068 | 57,067 |
| 22,577,000 | 4.4% | 7.8% | 6.6% |
| Group VI > 10K | 519 | 4,600 | 49,779 |
| 18,279,000 | 3.3% | 5.9% | 5.7% |
| Suburban Counties | 2,028 | 12,970 | 122,717 |
| 46,310,000 | 12.9% | 16.6% | 14.1% |
| Rural Counties | 1,024 | 5,709 | 44,217 |
| 22,903,000 | 6.5% | 7.3% | 5.1% |

| Population Group | Robbery | Bulgary |
|---|---|---|
| All 220,142,000 | 442,907 | 2,062,978 |
| Group I | | |
| 1 mil. | 120,847 | 228,388 |
| 20,922,000 | 27.3% | 11.1% |
| 500-999K | 50,939 | 147,987 |
| 11,084,000 | 11.5% | 7.2% |
| 250-499K | 69,436 | 214,878 |
| 13,825,000 | 15.7% | 10.4% |
| Group H 100-249K | 64,349 | 282,612 |
| 22,084,000 | 14.5% | 13.7% |
| Group HI 50-99K | 40,670 | 210,953 |
| 21,279,000 | 9.2% | 10.2% |
| Group IV 25-49K | 25,367 | 177,927 |
| 20,878,000 | 5.7% | 8.6% |
| Group V 10-24K | 18,153 | 164,427 |
| 22,577,000 | 4.1% | 8.0% |
| Group VI > 10K | 9,770 | 132,154 |
| 18,279,000 | 2.2% | 6.4% |
| Suburban Counties | 38,904 | 350,612 |
| 46,310,000 | 8.8% | 17.0% |
| Rural Counties | 4,472 | 153,040 |
| 22,903,000 | 1.0% | 7.4% |

Table 7.8 provides population data by various groupings and also provides UCR Index crime counts for these same groupings. Using these FBI data we are unable to replicate the 49% population figure. We come up with a total for the large population areas, those with at least 500,000 residents, of about 78,316,000. This was achieved by adding the population for Group I, I million or more (20,922,000), plus Group II, 500,000-999,999 (11,084,000), plus all of the suburban counties (46,310,000). Therefore, our calculations indicate that the two largest population areas, together with all the suburban counties equals about 35.6% of the United States population. Needless to say, this lack of congruence is problematic.

Next, we account for the percentage of UCR Index crimes that occur and are reported to the FBI for these three population groups that encompass densely populated areas. The results are as follows: UCR Index crime—35.1%; UCR violent Index crimes—42.6%; and UCR property Index crimes—34%. With the exception of a slightly disproportionate share of serious violent crimes, the largest population groups taken together account for basically the same amount of crime (34% to 35.1%) as their share of the population (35.6%) would suggest. Further, Table 9 indicates that there are no startling differentials in the extent to which the larger areas succeed in clearing the Index crimes by arrest. In fact, Table 7.9 generally indicates that it is the smaller areas that do a better job of arresting suspects regardless of the severity of the crime.

**Table 7.9. Percent of Offenses Cleared by Arrest by Population Group and Index Offense Type**

| Population Group | Total | Violent | Property | Murder/ Mans-laughter | Index |
|---|---|---|---|---|---|
| All 221,223,000 | 21.6 | 48.3 | 17.9 | 66.1 | |
| Group I | | | | | |
| 1 mil. + | 16.7 | 38.0 | 11.5 | 60.1 | 18,156,000 |
| 500-999K | 19.3 | 43.4 | 15.1 | 55.8 | 10,382,000 |
| 250-499K | 18.7 | 41.8 | 14.5 | 61.1 | 12,024,000 |
| Group II 100-249K | 21.2 | 47.9 | 17.5 | 69.1 | 19,631,000 |
| Group III 50-99K | 22.5 | 49.0 | 19.2 | 71.0 | 22,613,000 |
| Group IV 25-49K | 22.7 | 49.8 | 19.8 | 69.8 | 21,496,000 |
| Group V 10-24K | 26.0 | 55.3 | 23.2 | 71.9 | 23,519,000 |

| Population Group | Total | Violent | Property | Murder/ Mans- laughter | Index |
|---|---|---|---|---|---|
| Group VI > 10K | 24.8 | 61.3 | 21.5 | 74.9 | 19,329,000 |
| Suburban Counties | 21.7 | 54.9 | 17.4 | 68.8 | 50,186,000 |
| Rural Counties | 23.9 | 62.2 | 19.2 | 77.2 | 23,887,000 |
| **Population Group** | **Rape** | **Aggravated** | **Robbery** | **Burgalry** | **Assault** |
| All 221,223,000 | 50.8 | 58.5 | 26.3 | 13.8 | |
| Group I | | | | | |
| 1 mil. | + 46.0 | 50.3 | 20.5 | 9.4 | 18,156,000 |
| 500-999K | 55.2 | 55.7 | 25.3 | 13.2 | 10,382,000 |
| 250-499K | 50.3 | 52.2 | 24.2 | 10.6 | 12,024,000 |
| Group II 100-249K | 51.2 | 59.0 | 27.6 | 12.8 | 19,631,000 |
| Group III 50-99K | 46.8 | 59.7 | 27.7 | 13.7 | 22,613,000 |
| Group IV 25-49K | 44.6 | 58.2 | 29.5 | 13.5 | 21,496,000 |
| Group V 10-24K | 49.6 | 62.4 | 34.5 | 15.9 | 23,519,000 |
| Group VI > 10K | 55.2 | 66.7 | 36.3 | 17.0 | 19,329,000 |
| Suburban Counties | 54.9 | 62.7 | 29.6 | 18.6 | 50,186,000 |
| Rural Counties | 52.0 | 65.3 | 42.2 | 17.0 | 23,887,000 |

We come then to Effectiveness Test #3. As was the case previously with respect to law enforcement, the federal government's own data regarding prosecutors would suggest that DNA is not used by a majority of prosecutors' offices. There were population differentials showing greater use in the most populous areas. However, the extensive use by prosecutors serving the largest population groupings must be discounted by the fact that these offices handle only about one-third of the serious crime in the United States. The vast majority of crime, about 64%, occurs in smaller jurisdictions—jurisdictions served by prosecutor units that do not rely on DNA evidence to a great extent, either at trial or during plea negotiations, regardless of the severity of the crime being prosecuted. Our analysis of the available data leads us to one inescapable conclusion: DNA use is a big city or large county phenomenon, but the majority of crime is committed and prosecuted elsewhere.

## Cost-effectiveness

DNA mining has been shown above, at least anecdotally, to have the capability to match crime scene evidence with known offender samples. With respect to routine police and prosecutorial functions, however, DNA evidence is not being used extensively. Further, there appear to be distinct limitations to its capability to fight crime on any large scale, especially at the law enforcement and investigatory stages. Thus, it may seem unnecessary to consider cost-effectiveness. But, since DNA mining is here to stay, and its use appears likely to expand even further to include more and more donors (arrestees, convicted felons, perhaps even the general public) it is important to assess whether using DNA testing is cost-effective. At some point, consideration of allocating scarce law enforcement resources must be factored into the DNA mining equation, especially if successful "hits" remain few and far between. The alternative, of course, is that an overabundance of public expenditures might be allocated to DNA mining, even though it is not nearly as effective as has been claimed. There are several issues that warrant careful study in this regard. First, are DNA databases cost-effective right now? If not, will they be cost-effective at some point in the future? Even if such expenditures are the most efficient use of the resources allocated to crime fighting, do we really have sufficient resources to make the necessary and substantial fiscal outlays in order to capitalize on these new technologies?

The answers to all of these questions and more control the debate over whether allocating funds to these databases make economic sense. In one sense, however, it seems ludicrous to talk about cost-effectiveness at all when one reviews the anecdotal evidence of databases that have provided the only means for solving some crimes. What is the value of solving a brutal rape of two young sisters who were forced to watch as the perpetrator violently assaulted one and then the other? It seems that almost any amount of money could be justified as a reasonable expenditure. But, isolated successes are not the proper framework for examining the cost-effectiveness of DNA databases. The proper focus should be whether society is better off by spending money on these databases, which, as has been demonstrated above, may have a very low yield, as opposed to allocating the resources to other crime-fighting techniques.

As DNA laboratories all over the nation continue to upgrade existing technology, the capital investments will likely decline. Once these laboratories are fully equipped to handle DNA analysis on a large scale, the marginal cost of each analysis should decrease accordingly. It currently costs approximately $50 to construct a DNA profile from a sample. This year alone, the National Institute of Justice has begun distribution of $5 million in grants to speed DNA analysis technology. The National Commission on the Future of DNA Evidence has predicted that it would cost an estimated $22 million just to analyze the existing nationwide backlog of DNA samples. New York City Police Commissioner Howard Safir has proposed taking a DNA sample from "anyone who is arrested for anything." It has been estimated that this would cost New York City alone some $18.25 million per year. If this were expanded on a nationwide basis, it would result in about 15.3 million additional tests each year. When multiplied by the cost of each test, this additional expense would come to some $765 million annually (adjusted downward, of course, for offenders with multiple arrests).

These figures may seem daunting, but again, when compared to the costs of crime each year, and when compared to the efficacy of existing crime-fighting techniques, these expenditures may well be justified to some observers. It is also widely accepted that the costs of DNA testing will continue to decrease as technology improves Moreover, the increased use of DNA evidence could conceivably reduce the costs associated with traditional police investigations. Further, the utility of DNA as evidence should reduce backlogs in our court systems, and may serve to deter recidivism, thereby further justifying the expenditure on a cost-effectiveness basis. The use of DNA has been touted as one of the most effective tools our law enforcement agencies have ever had to fight crime. It is also claimed that DNA databases might diminish the cost of policing, investigating, and prosecuting criminal cases. These and other benefits make the expense seem more than worthwhile despite DNA mining's limited direct effect on crime-solving thus far.

It is difficult to locate precise figures on the cost of constructing and maintaining the current databases, and further, it remains unclear which category of offenders (or even people generally), and consequently, how many people, will have to contribute their DNA to the various databases. Notwithstanding the above limitations, we can offer a close examination concerning the cost-effectiveness of one of the newest databases to come on the scene: the Federal Convicted Offender DNA Database. The FBI has recently been authorized to implement a DNA database of data collected prior to release from a federal correctional institution from federal offenders convicted of an offense against a minor or a sexually violent offense. The FBI, in its budget request for fiscal year 2000, advised the House Committee on Appropriations that it would require $5,336,000 to implement the new federal database We are curious, if not very skeptical. Why will it cost such a substantial amount of money to add yet another database? Is there not a principle of "economy of scale" by which supplementing an already existing database (i.e., CODIS) could be accomplished without duplicating costs? Perhaps it is because there will be a considerable number of federal inmates subjected to the new program, thereby justifying the high cost.

In order to estimate the number of federal inmates who qualify for the new DNA database, we draw upon the federal government's own data sources. The Bureau of Justice Statistics has published two specialized reports on sex offenders: one on child victimizers and another on sex offenders generally. These reports virtually ignore sex offenders under federal custody, with the exception that there is one reference noting that about 1% of federal inmates in 1994 were serving time for a sex offense involving a victim who was a minor. Thus, not even the federal government's own specialized reports pay much attention to that segment of the sex offender population that is currently in federal custody. There are however, other sources from which to validate the small number of such offenders. The Federal Justice Statistics Database provides the following counts of sex offenders under federal custody: 1997—871 inmates (113 rapists and 758 other sex offenders); 1996—811 inmates (123 rapists and 688 other sex offenders); and 1995—731 inmates (120 rapists and 611 other sex offenders). Since the new database will likely collect inmates' DNA prior to release, the number of eligible inmates exiting the federal prison system each year becomes crucial to prospectively estimate the workload of the new database. The number of sex offenders exiting federal prison in recent years are: 1997—348 inmates (28 rapists and 320 other sex offenders); 1996—328 inmates (22 rapists and 306 other sex offenders); and 1995—249 inmates (29 rapists and 220 other sex offenders).

It should be very clear that, when compared to other offender groups, these numbers are extraordinarily small. For the last three years for which the above data are available, on average there were 811 sex offenders in federal custody, and 308 such offenders were released each year. According to the Federal Bureau of Prisons home page on the Internet, the official population count as of September 2, 1999 was 116,775 inmates. Using the Bureau of Justice Statistics 1% estimate, there would be 1,168 sex offenders in custody. Thus, we ask again, why such a substantial line item for a new database which will involve such a small number of offenders? Using the industry rate of $50 per DNA test from a blood sample, it would cost about $584,000 to test every federal inmate currently in custody, or alternatively, it would cost about $154,000 to test the 308 or so who would be expected to be released in any given year. One certainly wonders why the budget request asked for $5.3 million. In the final analysis, it would appear that the new Federal Convicted Offender DNA Database will hardly be cost-effective. We should all look forward to seeing whether such an extravagant budget is accompanied by significant crime reduction. Of course, if the analyses reported herein are any indication, we will all be very disappointed in the results.

It would seem that, given the questionable effectiveness of DNA in forensic settings, especially with respect to the value of DNA databases, public officials would be hesitant, or at least proceed slowly, in proposing costly DNA databases. However, as we will show in the following section, there has indeed been a rush to judgment that such databases are highly desirable, if not fundamentally necessary, in the fight against criminals.

## Future DNA Mining

Convicted sex-offenders? No problem here. They're all a bunch of sick recidivists anyway. All convicted felons? Sure, why not? Maybe they'll think twice next time. All people arrested? Well, it's a little tougher here, but sure, the cops are good guys and they can't arrest you without probable cause anyway. Everyone at birth or when we get our driver's license renewed? Well, maybe this is what is needed to end crime altogether. After all, if everybody knows that their DNA is on file and, consequently, that they will be caught every time for every crime, then no one would commit crimes. This hypothetical debate is happening in state legislatures everyday. This section of the paper is devoted to the future of DNA databases.

## A. What the Future May Hold

"Uncle Sam Wants You!" was the ramiliar slogan used to recruit young men for military service. Now, it might be more appropriate to say that "Uncle Sam Wants Your DNA!" On March 1, 1999, Attorney General Janet Reno asked a national commission to study the legality of taking DNA samples from everyone arrested instead of just the convicted sex offenders and violent felons currently permitted by law. The committee's recommendations are not available yet, but it would be a safe bet to predict at least some expansion over the current status quo. Louisiana is the most striking example of the expansionist tendencies these DNA databases have, but New York City, North Carolina, and England are also highly illustrative. At least one poll shows that people are split almost evenly on the issue of whether DNA samples should be taken from everyone arrested. Of 9,751 responses, 54.4% answered yes, 41.6% said no, and 3.9% were not sure. With such a slim margin of support, one may properly ask whether any policy should be enacted. For issues as serious as those DNA databases present, it might be appropriate to only enact legislation that captures a broader consensus and approval by the populace.

Aside from these calls for moves to the very limit of permissible sampling, other states are enacting laws or have enacted laws expanding their DNA database "trigger mechanisms." Colorado passed a bill mandating DNA tests for certain felonies involving bodily harm as a condition of the felon's parole. Many of the states that currently restrict their databases to sex offenders are considering expansion to bring them more in line with the majority of jurisdictions which allow DNA profiling for some or all felony convictions.

### 1. Possible Scenarios

Although, in theory, there would seem to be an infinite number of possible DNA database systems which can be developed to take advantage of DNA technology and provide fertile ground for DNA mining, in reality, there are really just seven likely outcomes.

#### a. No DNA Databases at All

The option of no DNA databases at all seems highly unlikely now that every state in the union has a DNA database system either in operation or under development. Given the alleged advantages that DNA has for crime-solving activities, and its secondary effects on crime prevention, one would not be too adventuresome to predict that these databases are here to stay whether we like it or not.

**b. Just Sex Offenders**

This was the early trend, but as advancements in DNA technology continue – and they will – it remains to be seen whether DNA databases will continue to be limited to just these types of offenders. If society's goal is to prevent and catch this type of criminal activity, then it serves its purpose well. However, if society feels the need to expand its crime-solving abilities, these databases will be broadened accordingly.

**c. All Persons Convicted of Felony Offenses**

This trend is becoming more popular, as it meshes nicely with the "get tough" on crime mentality among the general population. Further, it catches some criminals who would not otherwise be included under the sex offender-based programs. For example, if a criminal was to rape and kill his victim, this is likely to be classified as a murder, and not a sex offense. Thus, the offender would not be subject to inclusion in the database, even though this is precisely the type of heinous crime that the sex-offender database was designed to prevent.

**d. All Persons Convicted**

Under such a system, every conviction would result in a DNA database record. This might have a real and substantial impact on a person's desire to commit future criminal acts. However, it is not clear that this would be worth the considerable expense and effort such an undertaking would entail. As technology becomes cheaper, faster, and more user friendly, it might make such programs less cost-prohibitive. One has to wonder how useful the inclusion of someone who was arrested for a so-called "victimless" crime, such as gambling, would be. It seems doubtful that a person who was arrested for betting $50 on the Bulls to beat the Lakers would significantly improve the capability of the database to identify serial rapists. On the other hand, that bettor may just be the one person the police were looking for.

**e. All Persons Arrested for Felony Offenses**

The problem here, and with any other system which does not require a conviction, is that it has a tremendous potential for abuse.

Where there is no judicial check on the intake process, it is difficult to convince those who view privacy as one of our most cherished rights that DNA testing is appropriate in this situation. This is particularly true today as we have increasingly looked to courts, and not legislatures, as the primary guardians of our liberties.

**f. All Persons Arrested**

Soon to take effect in Louisiana, and proposed in New York City, this is the current outer-limit of DNA databases. To date, no system has pushed farther than this. This makes sense, but perhaps only because the next alternative, total population inclusion, is a daunting prospect indeed. Thus, this appears to be our current limit for the use of DNA databases. From one perspective, taking DNA from all arrestees may enable police to solve crimes where only biological evidence was left. A criminal who is arrested for one crime may be wanted in connection with other crimes, but solely on the basis of DNA evidence recovered from a crime scene. This would allow law enforcement authorities to clear many unsolved crimes, which is obviously a tremendous benefit to society. From another perspective, our criminal justice system is based on the concept that we are innocent until proven guilty. Thus, it seems perhaps a little too intrusive to take DNA samples simply on the basis of suspected criminal activity.

**g. Total Population Inclusion**

The final alternative is a database with a record of the DNA profiles of all people. First, the logistical problems here are enormous. How should this category be defined? Every person in the state? A national database with every American citizen? Taken at birth? What about those already born? Does everybody have to give a sample at the local police station? Where do we get the funding for this? Who would be in charge of it? How do we deal with illegal aliens? What about resident aliens? Tourists? Obviously, we cannot force other countries to maintain such a database, nor could we force them to share the data with our law enforcement agencies. The international aspects alone present almost insurmountable obstacles to the successful implementation of such a system.

Aside from the practical obstacles, how would we feel about such a database? Assume not only that such a database is possible, but further, that it is entirely cost-free to develop. No lost time or productivity to give and collect samples; no funds required for the machines to do the analysis; no funds needed to pay the technicians to organize the data. Would we still want it, even if it were free? Reasonable people might disagree, even with the completely fictional assumption that it would cost nothing to create.

These various scenarios follow a clear-cut progression. A sex-offender-only database seems virtually unobjectionable. The conviction-based databases seem a little more expansive, but still they do not seem particularly troubling. After all, a judge or jury found the accused guilty. That should be enough. The arrest-based systems are the current thresholds, but here, we have no assurance of guilt, only suspicion. With the total population database, even if one takes cost out of the equation, it still seems a little too futuristic. Something about it just seems contrary to our notions of individual autonomy, and our sense of personal privacy. These are the concerns that DNA databases ultimately present. There are no easy answers. In fact, there are very few easy questions.

## B. Do We Want To Go There?

"Better living through chemistry" was a popular slogan many years ago describing how science could make our lives better. It may be even more appropriate to use it today. Thankfully, the notion of privacy has not been lost in the debate. Civil libertarians are answering the call, and state legislatures are not blind to the issues themselves. Aside from pure privacy concerns, there are legitimate fears about misuse of these databases. Other concerns such as the integrity of the DNA analysis process, and whether we should now revisit heretofore accepted rules regarding things such as statutes of limitations are key issues which must be resolved.

Once the DNA is on file, concerns arise as to what other uses it may be put to. This has been called "function creep" by one leading commentator. To allay these concerns, almost every DNA database statute on the books contains provisions to avoid such improper usage. Rhode Island's DNA database act provides: "DNA samples and DNA records collected under this chapter for the purpose of obtaining information about physical characteristics, traits or predispositions for disease." Texas has an explicit prohibition against the use of genetic data by insurance companies for illegitimate purposes. Article 21.73 of the Texas Insurance Code states: "A group health benefit plan issuer may not use genetic information to reject, deny, limit, cancel, refuse to renew, increase the premiums for, or otherwise adversely affect eligibility for or coverage under the group health benefit plan." The statute further holds refusal to submit to a genetic test may also not be used in the same manner. Moreover, such prohibitions do not apply only to people already born. The statute also mandates that "no issuer of a group health benefit plan shall use genetic information to coerce or compel a pregnant woman to have an induced abortion."

This debate will be carried out for years to come, and it is reasonable to expect that as science continues to progress, even more invasive and intrusive issues will be considered. A watchful eye should ensure that we can embrace the benefits of this technology, but without giving up too much in exchange. As Part III.C focused on the monetary costs we incur to pay for these databases, we must also focus on the costs in terms of sacrificing personal liberties as well.

## Existing State of the Law

Given the proliferation of DNA databases and the proposals for wide expansion, it is desirable to review briefly the statutes and regulations already "on the books" as of the time of this writing. In so doing, we can examine the extent to which such statutes are uniform, if not identical, thus providing, regardless of jurisdiction, standardized procedures and policies concerning database information. On the other hand, anomalous provisions should be identified and their possible implications raised.

In addition to the FBI's DNA database, every state in the union has established a DNA database in one form or another. This is likely due, at least in part, to the availability of federal funding for such endeavors. As a condition of the federal government's support, each state's database must meet certain criteria. For example, in order to receive a grant, the state must require that DNA samples be obtained from "each person convicted of a felony of a sexual nature." Further, each state must require that its standards for the processing and analysis of such specimens are consistent with those established by the Director of the FBI. These various statutes are generally the same for the most part, but there is substantial variance as to what groups of people are forced to provide samples for inclusion in a database. To provide a useful analytical framework, the Texas DNA Database System will be reviewed in depth, followed by a discussion of other states' approaches, and then a discussion of anomalous provisions of particular statutes will complete this section.

## A. The Texas DNA Database System

The legislation creating Texas' DNA database system was enacted in 1995. Chapter 411 of the Texas Government Code covers the Texas Department of Public Safety, which has the responsibility for the DNA database.

**1. Section 411.142: "DNA Database"**

The Director shall "record DNA data and establish and maintain a computerized database that serves as the central depository in the state for DNA records. A "DNA record" means the results of a forensic DNA analysis performed by a DNA laboratory and, if known, the name of the person who is the subject of the analysis." The director may "receive, analyze, store, and destroy a record, blood sample, or other specimen...." The DNA database must "be capable of classifying, matching, and storing the results of analyses of DNA and other biological molecules." The DNA database must also "be compatible with the national DNA identification system (CODIS) used by the FBI to the extent required by the FBI to permit the useful exchange and storage of DNA records or other information derived from those records." The department shall "establish standards for DNA analysis ... that meet or exceed the current standards for quality assurance ... issued by the FBI."

**2. Section 411.143: "Purposes"**

The "principal purpose of the DNA database is to assist federal, state, or local criminal justice or law enforcement agencies in the investigation or prosecution of sex-related offenses or other offenses in which biological evidence is recovered.'" In criminal cases, the "purposes of the DNA database are only for use in the investigation of an offense, the exclusion or identification of suspects and the prosecution of the case." Other purposes of the database include:

> *1) assisting in the recovery or identification of human remains from a disaster or for humanitarian purposes; 2) assisting in the identification*

*of living or deceased missing persons; and 3) if personal identifying information is removed: A) establishing a population statistics database;B) assisting in identification research and protocol development; and C) assisting in database or DNA laboratory quality control.*

3. Section 411.144: "Regulation of DNA Laboratories; Penalties" According to section 411.144:

*The director shall establish procedures for a DNA laboratory or criminal justice or law enforcement agency in the collection, preservation, shipment, analysis, and use of a blood sample or other specimen ... in a manner that permits the exchange of DNA evidence between DNA laboratories and the use of the evidence in a criminal case.*

The director may "at any reasonable time enter and inspect the premises or audit the procedures of any DNA laboratory that provides DNA records or DNA forensic analyses to the department...." Eventually, there are plans for eight separate DNA labs which would feed data to the state's DNA database, although none are presently operational.

4. Section 411.146: "Blood Samples or Other Specimens"

A person "collecting a blood sample or other specimen may not be held liable in any civil or criminal action if the person collects the sample or specimen in a reasonable manner according to generally accepted medical or other professional practices." A DNA laboratory may "analyze a blood sample collected under this section or other DNA specimen only: to type the genetic markers contained in the sample or specimen; for criminal justice and law enforcement purposes; or for other purposes described by this sub-chapter." "If possible, a second DNA specimen must be obtained from a suspect in a criminal investigation if forensic DNA evidence is necessary for use as substantive evidence in the prosecution of a case."

5. Section 411.147: "Access to DNA Database Information"

The director shall "establish procedures: to prevent unauthorized access to the DNA database; and to release DNA records, specimens, or analyses from the DNA database." The department may:

*Release a DNA sample, analysis, or record only: to a criminal justice agency for law enforcement identification purposes; for a*

> *judicial proceeding, if otherwise admissible under law; for criminal defense purposes to a defendant, if related to the case in which the defendant is charged; or if personally identifiable information is removed, for: a population statistics database; identification research and protocol development; or quality control.*

The director shall "maintain a record of requests made under this section."

6. Section 411.148: "DNA Records of Certain Inmates"; and Section 411.150: "DNA Records of Certain Juveniles"

The Texas DNA database can contain records that originate from eight separate and distinct sources. For our purposes, however, the only two sources that are relevant to this paper are: a person described by section 411.148 or 411.150, and a biological specimen that is legally obtained in the investigation of a crime, regardless of origin. Sections 411.148 and 411.150 delineate the offenses for which an adult or a juvenile may be required to submit a DNA sample for inclusion into the database. Both sections are essentially identical and cover the same offenses. Under section 411.148, an adult can be compelled to submit a biological specimen essentially if that person commits a sex-related offense, another felony offense with the intent to commit a sex-related offense, murder, or aggravated assault. Also, under both section 411.148 and section 411.150, an inmate can be forced to give a sample by order of a court. This would appear to give courts free reign over the DNA database, but there is no evidence that any such orders have been given to date. Further, it appears that under the Texas statute, an inmate can be forced to give a sample if the inmate is serving a sentence for any offense and has been previously convicted of a sex-related offense.

Under section 411.148, an inmate "may not be held past a statutory release date if the inmate fails or refuses to provide a blood sample or other specimen under this section." However, "[a] penal institution may take other lawful administrative action against the inmate." Section 411.150, dealing with juveniles, is substantially similar, although unlike section 411.148, the state is not prohibited from holding a juvenile past his release date for failure to provide a sample. Aside from this noteworthy omission, the sections seem to track each other very closely.

7. Section 411.151: "Expunction of DNA Records"

The director shall "expunge a DNA record of a person from the DNA database if the person: notifies the director in writing that the

DNA record has been ordered to be expunged and; provides the director with a certified copy of the court order that expunges the DNA record." A person may "petition for the expunction of a DNA record ... if the person is entitled to the expunction of records relating to the offense to which the DNA record is related...."

8. Section 411.153: "Confidentiality of DNA Records"

> *"A DNA record stored in the DNA database is confidential and is not subject to disclosure under the open records law...." A person commits an offense if the person knowingly discloses information in a DNA record ... except as authorized by this chapter. An offense under this subsection is a misdemeanor punishable by: a fine of not more than $1,000.; confinement in the county jail for not more than six months; or both the fine and confinement.*

9. Section 411.154: "Enforcement by Court Order" Section 411.154 states:

> *The court may issue an order requiring a person: to act in compliance with this subchapter or a rule adopted under this subchapter; to refrain from acting in violation of this subchapter or a rule adopted under this subchapter; to give a blood sample or other specimen; or if the person has already given a blood sample or other specimen, to give another sample if good cause is shown.*

Orders issued under this section are appealable as criminal matters and are to be reviewed under an abuse of discretion standard. As of this writing, the Texas DNA database has produced a total of four "cold hits." Currently, the Texas DNA database has approximately 12,000 samples on file. There is presently a backlog of approximately 7,000 samples which have been received, but not yet analyzed. The analysis process of DNA specimens takes about two to three days, using current technology. To date, the Texas DNA database has cost approximately $1.8 million, and it is currently staffed by eleven individuals. The Texas legislature has recently expanded sections 411.148 and 411.150 to include the offenses of murder, aggravated assault, and certain burglaries. Interestingly, there is a difference of opinion in the Department of Public Safety as to exactly how far the Texas DNA database should be expanded. The CODIS lab supervisor is of the opinion that just adding murder, certain assaultive offenses, and burglary is all that is necessary, while one DNA crime lab technician we interviewed would like to see all arrests be included. While this finding is not entirely surprising, it is indicative of the contrasting views related to the expansion of these databases.

The Texas DNA database statute is fairly representative of a number of the DNA database provisions of other states. Now that a review of the Texas statute is complete, it can be compared and contrasted with those of other states.

## B. Other Jurisdictions Generally

At the outset, it is important to note that there are states that have "trigger mechanisms" which are narrower than the Texas version, and there are states whose statutes have a much broader scope than Texas. However, the Texas statute accurately depicts the nature and scope of its counterparts in many other jurisdictions. While most of these statutes contain many of the same general features, the critical divergence becomes apparent when the analysis focuses on a single question: what conduct will result in a person being forced to submit a DNA sample for inclusion in a database? This inquiry allows the various statutes to be segregated and categorized, much like DNA samples themselves. Based on the variety among them, it is hazardous to generalize about all other states. However, the databases of other states can fairly be divided into two broad groups. These groups are: states which include DNA only from persons with convictions for felony sexual offenses; and states which include DNA from persons with convictions for some or all of the state's felony offenses. The reason that these two classifications were chosen is that it enables one to understand the expansionist tendencies regarding these databases. Typically, they were set up in order to target sex offenders only. However, they have tended to gradually expand to encompass ever broader segments of the population.

There are two states, however, which flatly do not fall within this rubric. Louisiana and Iowa have statutes in a class all by themselves. Louisiana currently has the most expansive statute on the books. Beginning on September 1, 1999, all persons "arrested for a felony sex offense or other specified offense shall have a DNA sample drawn or taken at the same time he is fingerprinted pursuant to the booking procedure." As the statute now reads, a person can be forced to submit a DNA sample merely for being arrested for a simple assault. Under Louisiana law, a person who commits a simple assault "shall be fined not more than two hundred dollars, or imprisoned for not more than ninety days, or both."

Iowa's DNA database system is different than every other state as well. The legislature deferred to the law enforcement divisions of the state, and allowed almost total freedom for the attorney general, courts, and parole boards to define what offenses should trigger inclusion in the DNA database. The Attorney General shall "adopt rules ... for the purpose of classifying felonies and indictable misdemeanors which shall require the offender to submit a physical specimen for DNA profiling. Factors to be considered shall include the deterrent effect of DNA profiling, the likelihood of repeated violations, and the seriousness of the offense."

The above emphasis on the "trigger mechanisms" might overstate the similarities a bit, however, because there are significant differences among the statutes exclusive of the issue as to what offenses will result in someone's inclusion in a DNA database. For example, some states require the offender to pay a fee for the costs of running DNA analysis on their sample, while others do not. In addition, some states expressly provide a cause of action against persons who violate the confidentiality and anti-tampering provisions of statutes, while others do not. A few of these issues are worth viewing in the comparative context as well.

## C. Database Issues

While the list detailed below is by no means exhaustive, it does raise a number of important topics relating to the mechanics of these databases. Consider the following questions:

1. Who has access to these databases?

   Most states restrict access to law enforcement agencies, court proceedings, and to a defendant in connection with the defense of the charge that gave rise to the DNA sample.

2. What security measures are commonly utilized?

   Most states have penalties designed to combat unauthorized use and tampering.

3. Does the statute grant immunity to those who draw samples?

Over thirty states allow some form of immunity, both from criminal and civil liability arising out of the taking of a sample. However, the jurisdictions are about evenly split on whether the immunity covers negligence.

4. Is reasonable force allowed to obtain a sample?

   Missouri is one of only eight states that allow the use of reasonable or necessary force to obtain a sample from uncooperative subjects.

5. Does the statute allow the charging of fees for the expense of analyzing the DNA?

   Like more than one dozen states, South Carolina allows for costs to be assessed against an offender.

6. Does the statute allow for expunction of DNA records?

   Over three dozen states have included expunction provisions expressly related to the DNA database.

The statutes as a whole have many similarities, but there are key distinctions among them. The fact that these statutes are subject to almost constant revision presents a rather large obstacle to a comprehensive survey, and due to the rapidly changing nature of these laws, any comparative study is really more like a photograph that freezes a moment in time.

## D. Anomalous Provisions

Despite a general congruence among the different statutes, there are several unique provisions that are worth mentioning. A small number of these anomalies are presented here for the reader's consideration. First, Pennsylvania's statute ordered the commissioner of the State Police to publish a notice when the database was fully operational. Perhaps this was a legislative attempt to circumvent any Lambert claims (a claim which arises when a person argues a lack of knowledge of a law's applicability). Presumably, a person subject to inclusion in the database would not be successful in challenging the law on the grounds that it was unknown to the general public, or that it was not easily discoverable. By publicly announcing the readiness of the DNA database system, a challenger would have a difficult task to prevail on a Lambert claim. Second, South Carolina's fee recoupment provision involves a $250 charge which cannot be waived by a court, and a prisoner may not be released from prison on parole or even after the completion of his sentence until the fee is paid. Third, Virginia, which had the first DNA database in the country, now requires DNA samples for every person convicted of a felony. However, unlawful use of the database information can be a felony. Thus, by misusing the DNA database, a person could easily find themselves the newest member of the database.

## Conclusion

We better off living in a world where our most basic and singularly unique characteristics are on file, serving as a constant shadow over our daily lives? The question suggests a brooding omnipresence of big government that makes us uncomfortable. Yet, if DNA databases could be proven to be of unparalleled value in fighting crime, then the answer might be yes, other concerns notwithstanding. In this vein, we have no quarrel whatsoever with the earliest and most basic of DNA applications in the criminal law: known suspect testing and post-conviction relief. We agree that there is unparalleled value to the use of DNA testing to match a particular suspect's DNA with that extracted from trace evidence left behind at a crime scene. Similarly, can there be a more justice-oriented application for DNA than to use its exculpatory capabilities to exonerate persons who were wrongfully accused and convicted? Moreover, aside from the obvious moral issues, a wrongful conviction serves no purpose, and consequently is of no value in our criminal justice system. Each of these DNA applications should be used as extensively as possible, not only as effective crime control measures, but more importantly, as definitive tests of whether an accused, or even a previously convicted person, is actually innocent.

Beyond these two applications of DNA testing per se, however, our inquiry concerning the spreading craze over DNA databases as a crime control measure does not offer similar support. In fact, the analysis reported here of the best available government crime data raises serious concerns that DNA databases are proliferating and becoming ever more inclusive, and the costs associated with collecting, testing, and storing the information are rising into the hundreds of millions of dollars. These developments are occurring all across the country despite the absence of convincing evidence that the DNA mining process will strike gold as proponents have claimed.

We have demonstrated that DNA databases will not be greatly successful in increasing the extent to which police solve the vast majority of the seven FBI Index crimes. This was shown to be the case because the vast majority of Index crimes are property offenses, and this type of offense does not carry a high potential for beneficial DNA testing because the usual property offense crime scene is not likely to contain the perpetrators' DNA, and even if it does, such evidence will seldom be looked for, let alone collected and tested for comparison with the DNA databases. We also showed that DNA mining will be even less beneficial for solving the vast majority of Non-Index crimes. Again, we argued that the millions of less serious Non-Index crime scenes hardly ever contain DNA-related evidence. We further argued that even if such crime scenes did contain DNA evidence, local law enforcement hardly has the necessary resources to treat these offenses as though they deserved the intensive crime scene effort that is usually reserved for serious violent crimes against the person. In this regard, we noted that it is often difficult enough to convince the police to dust for fingerprints at a residential burglary, because the police know that their search will likely be futile. Imagine, therefore, trying to convince police to search the crime scene (usually outside) of a robbery for such evidence as the perpetrator's hair, tissue, or other residual evidence.

We also considered the value of DNA database evidence to the prosecutorial function. We found that DNA evidence is not used by a majority of prosecutors' offices across the country. We did find that "big city" prosecutors relied heavily on DNA evidence both at trial and during plea negotiations. However, this extensive use pattern in the most populous areas must be discounted by the fact that these offices handle only about one-third of the serious crime in the United States. That is, the vast majority of crime, about 64%, occurs in smaller jurisdictions, which are served by prosecutor units that do not rely heavily on DNA evidence, either at trial or during plea negotiations, regardless of the severity of the crime being prosecuted. We found that prosecutorial use of DNA is clearly a big city or large county phenomenon, but the majority of crime is committed and prosecuted elsewhere, thus diminishing the value of DNA as a crime control measure at the macro level. Further, we also found that even among the big city prosecutors, DNA evidence use was restricted to mostly homicides, rapes, and other very serious offenses.

We also considered the cost-effectiveness of DNA databases. Here the results were quite convincing. At present, the DNA extraction process is a highly expensive and time-consuming process when considered in the aggregate. The costs associated with increased testing, especially the increased testing necessitated by the more "inclusive" DNA database proposals are astronomical when compared to the expected crime level benefits associated with the databases. For example, we provided an examination of the soon-to-be-launched Federal Convicted Offender DNA Database and found that, while it will cost $5,335,000 the first year, the current total population of offenders who will be included in the database numbers about 1,200 inmates. This translates to some $4,445 per inmate. If the costs of analyzing a sample are only $50, then we must ask the question: Where is the rest of this money supposed to go? Clearly, a legitimate question remains: Are DNA databases and their direct and more indirect or diffuse costs the most effective way to spend scarce criminal justice resources? At this point, the answer must be no.

Last, we considered various future scenarios and proposals for various DNA database configurations. These proposals include very specific eligibility criteria, like persons convicted of sex offenses or violent crimes, but they also include more expansive criteria like persons arrested, the general public, or even newborns as has been endorsed by Mayor Giuliani of New York City. We argued that there are actual financial costs, as well as ethical and civil liberty costs associated with these ever expanding DNA database proposals. The essential point is clear. The only reason to bear the manifold costs associated with DNA databases is if, and only if, it can be shown that a particular database configuration will be demonstratively successful in solving crimes or easing prosecutions which would not be possible otherwise. Thus, what good are the more expansive (and expensive) database definitions like, "all persons arrested for anything?" Our response is: "Not very good." The only databases that appear worthy of serious consideration are those which focus on specific categories of criminals, like violent offenders and sexual predators. It remains open to debate whether such databases should contain just convicted persons or arrested persons as well. However, the restriction to particular offense types appears to be highly justified on the basis of the demonstrative yield owing to crime scene issues and the well recognized recidivism potential of violent offenders and sexual predators. We should all look with great skepticism at proposals for more inclusive eligibility. Further, we should demand that the proponents of wider inclusion prove scientifically and unequivocally the crime level value of the more inclusive databases. In proposing that DNA be taken from every person arrested in New York City, the Police Commissioner, Howard Safir has stated that, "A professional burglar or car thief, for instance, might leave skin or sweat cells when forcing open a door or window. As long as there is one cell, you could use this to help solve just about any crime you could think of."

Statements like this not only strain credibility, but give us all reason to pause and wonder whether public officials have succumbed to a fatal case of DNA fever.

There are ever present dangers that the public will be swept up in the same euphoria over DNA mining which proponents, like Police Commissioner Safir, seem to believe is the best thing to hit law enforcement since fingerprinting. Meanwhile, the implications for privacy and other personal liberty issues, plus the huge potential for abuse, should encourage us to keep looking over Big Brother's shoulder even as he watches us.

## DNA Databases for 21st Century

The greatest dangers to liberty lurk in insidious encroachment by men of zeal, well-meaning but without understanding. (1) Following the highly publicized trial of O.J. Simpson, DNA evidence has become a common feature in American criminal trials. As a forensic tool, DNA analysis was initially used to link an already known suspect to a particular crime scene. In these situations, law enforcement officers obtain DNA from the suspect and compare it to DNA recovered from the crime scene. While this direct analysis of a known suspect's DNA is still widespread, a new forensic use has developed. With the development of DNA databases, states are obtaining DNA samples from people convicted of specific crimes and then storing the resulting DNA "fingerprint" in a computer database.(2) When a crime occurs in which DNA evidence is recovered, police compare that sample against every previously obtained DNA fingerprint in the database, hoping to find a match, or "cold hit." This procedure, and DNA databases generally, allows law enforcement to use DNA in the initial stage of an investigation to identify a suspect.

The potential use of DNA to identify the offender in every crime where DNA evidence is recovered has caused some commentators to christen DNA fingerprinting as "the single biggest advancement in forensic science."(3) Optimal effectiveness, however, would require a universal DNA database that contains DNA fingerprint of every citizen, otherwise potential matches would be missed.(4) Some argue that a universal DNA database, in addition to solving crimes, would deter crimes because people would be unlikely to commit crimes knowing that the likelihood of being conclusively identified through a quick search in a DNA database was nearly one-hundred percent.(5) This Note discusses the efficacy of developing a universal DNA database by asking a larger question: how far are we as a society prepared to go in allowing the government to impede our liberty and our privacy in the name of crime prevention. Is there a limit, constitutionally or theoretically, to what the government can do when it is acting in its role as the preventive state (6) Have we, as a citizenry, become so terrified of crime that we are willing to suspend our concept of ourselves as a free and autonomous population? Have we as a citizenry, become so terrified of crime that we are willing to allow the government to infringe on every person's freedom by requiring citizens to provide DNA samples?

## Overview of DNA and DNA Fingerprinting

To understand why DNA is such an effective forensic identification tool, it is necessary to first discuss the DNA molecule itself. Part A briefly describes the DNA molecule, and explains why it is such a powerful means of identification. Part B describes the most common technique of DNA fingerprinting for forensic identification purposes, Restriction Fragment Length Polymorphism analysis.

## A. The DNA Molecule

Deoxyribonucleic acid is a long chemical strand found in the nucleus of almost every cell in living organisms.(7) The DNA molecule is essentially the "blueprint" for the physical make-up of each individual; it determines every physical characteristic from height to eye color.(8) A DNA molecule consists of two twisted strands of alternating units of phosphates and sugars connected, like rungs on a ladder, by complementary nucleotide base pairs.(9) Among the four nucleotide bases, Adenine (A), Thymine (T), Cytosine (C) and Guanine (G), A only links with T, and C only links with G. DNA is found on the twenty-three chromosomes a person inherits from each parent. Since these DNA sequences code for the characteristics that make us human, i.e., two hands, two eyes, two legs, etc., approximately ninety-nine percent of an individual's DNA is identical to all other humans. However, the remaining one percent of the DNA sequences vary from person to person. Within these sections of DNA, sequences of base pairs are often repeated hundreds or even thousands of times. Such sequences are called Variable Number of Tandem Repeats ("VNTRs"). Because the pattern of VNTRs is different for everyone, examination of these polymorphic sections allow an individual to be identified.

## B. DNA Fingerprinting

While there are at least two other types of DNA analysis that may be used to identify individuals, RFLP is currently the most commonly employed. The theory behind RFLP analysis is that by examining the length of VNTRs at various sites, or "loci," one can determine whether a particular sample of DNA came from a known source.

In RFLP analysis, first DNA is extracted either from an individual, or from evidence, such as blood or skin, obtained at a crime scene. The extracted DNA is then digested by an enzyme that cuts the chain wherever it contacts a particular known sequence. The result is numerous DNA fragments of varying lengths. Because the length of VNTRs in individuals varies, the enzyme will cut at different lengths for different people. The fragmented DNA is then placed on a slab of gelatin-like material known as agarose gel. An electric current is applied to the gel which causes the fragments to separate and move across the gel at varying speeds, depending on the length of the fragment. Shorter, lighter fragments travel further across the gel than longer, heavier fragments. These fragments are then transferred in their exact positions from the gel to a nylon membrane. Radioactive markers are then applied to the DNA fragments on the membrane and exposed to a sheet of film. The result is an autorad, the exposed films depiction of the banding pattern of the DNA, which creates a UPC-like bar code as found on items in retail stores. This banding pattern is the DNA fingerprint that can be digitalized and stored in a database for future use.

The DNA fingerprint is then compared to a known sample of DNA. If the DNA was from the same source, fragments of the same length should appear as bands on the same place on the autorad. A visual match is verified by a computer measurement of the digitized pattern which determines the two samples to be a match if they fall within a specified match window of measurement variation. Once a match has been identified, the examiner must determine the statistical probability that this particular profile occurs in a particular population. Generally this is done through the "product rule" in which the probability that one person in a particular population would display this banding pattern in one probe is multiplied by the probability that another person in the same population would display the same pattern for a second probe. In a criminal trial, the result might be explained by an expert in statistical probabilities as "the probability of selecting at random from the population an unrelated individual having a DNA profile matching the [defendant's] is approximately 1 in 200,000 in African Americans, 1 in 200,000 in Whites, and 1 in 100,000 in Hispanics." It is important to note that the only information obtainable about an individual genetic sequence from a DNA fingerprint created through RFLP analysis is whether two samples originated from the same person. The DNA fingerprint itself does not contain any information about the individual's eye color, race, sex, or risk of genetic disease.

## Offender DNA Databases and a Universal Dna Database

The primary concern of this Note is the development of a potential universal DNA database containing DNA fingerprints from every individual, to understand the constitutional and policy implications of a universal database it is first necessary to consider existing DNA databases presently limited to criminal offenders.

## A. Offender DNA Databases

With the increased use of DNA evidence in criminal trials to link the defendant to the crime scene, a movement to use DNA in the investigative stage to identify potential suspects has gained momentum. England was the first country to develop a large-scale database for this purpose. To identify as many suspects as possible, England requires samples to be taken from anyone charged with a "recordable offense." While the United States has progressed more slowly in creating DNA databases for criminal offenders, the passage of the DNA Identification Act of 1994 spurred the movement by creating a national offender DNA database administered by the FBI and by offering financial incentives for states to create their own offender DNA databases. Today, all fifty states have legislation requiring specific classes of convicted offenders to provide DNA fingerprints for inclusion in both their state database and the FBI's Combined DNA Information System (CODIS). The DNA sample is usually obtained through a blood sample provided by the offender. Most state statutes authorize the establishment of offender DNA databases as a tool to solve crimes where few other investigative leads are available. Increasingly, statutes are amended to allow for broader use of DNA fingerprints where such advanced identification techniques could help identify missing persons or the badly damaged bodies of victims of natural or other massive disasters. Initially, state statutes only required DNA fingerprints for inclusion in the database from offenders convicted of a small class of crimes such as homicide and sexual assault. The offenses were limited to such crimes because the likelihood of an offender leaving his DNA behind in the form of semen or blood was higher than in other crimes.

As technology now allows for extraction of DNA from smaller, more common bits of evidence such as hair and skin cells, the list of enumerated offenses that states are willing to include in their databases is expanding. Currently, at least four states require samples from anyone convicted of any felony. Louisiana has gone so far as to require samples from individuals merely arrested for sex offenses. Additionally, Arizona, Kansas, and Oregon require samples to be taken from juveniles found delinquent of certain sex crimes for use in investigations years after they reach the age of majority.

In most states, offenders are required to submit a sample while they are in prison. or before they are released as a condition of their parole or probation. Retroactive application to those currently serving their parole time often requires the submission of samples from individuals convicted in another state but whose parole or probation supervision is transferred to the current state. At times, submission of a sample is a condition of an offender's probation even if prison time is not required. In at least one state, forcible extraction is allowed if the prisoner refuses to cooperate. In Illinois, as of January 1, 2000, any deliberate act to delay or impede the collection of a sample from an individual covered under the statute is punishable as a Class A misdemeanor.

While some states explicitly allow for the digitized DNA fingerprint to be forwarded to the FBI and entered into the national database CODIS, others only require their databases and procedures be compatible with CODIS and available to outside law enforcement agencies. As of October 1998, CODIS contained over 250,000 DNA fingerprints and was responsible for over 400 matches between known offender DNA fingerprints and DNA left at crime scenes.

Access to both state and national databases is usually strictly limited to agents involved in the criminal justice system. Use of the DNA fingerprints as an identification device is limited to purposes listed in the statutes, generally law enforcement and identification of missing persons. Although most states punish misuse of the DNA database as a misdemeanor, others have elevated such acts to the level of a felony. Expungement of the DNA fingerprint from the offender DNA database following a subsequent reversal and dismissal of the conviction is allowed in an increasing number of states.

## B. A Universal DNA Database

As effective as existing DNA databases already are, they are by no means as powerful a resource as they could be. Existing DNA databases are currently limited to containing only the DNA fingerprints of offenders convicted of specific, enumerated crimes or of persons arrested because the police had probable cause to believe they committed such crimes. While states are expanding both the types of crimes the commission of which will lead to a sample being taken and the stage in the criminal process at which a sample can be required, the coverage of existing DNA databases is limited to only a small portion of the overall population of the United States. Additionally, states are finding themselves faced with a growing number of crime scene samples from unsolved cases without the laboratory funding or manpower to analyze and enter a DNA fingerprint into an offender database. Because of this backlog, not only are first-time offenders being missed, but crimes may be occurring which could be prevented had the DNA from a prior crime perpetrated by the same individual been analyzed and entered into the database in a timely manner. Despite the backlog, Virginia and Florida have increased the size of their databases to the point where they are both experiencing more than one "cold hit" a week. The deficiencies of current databases have led law enforcement officials to increasingly resort to "dragnet" procedures that target a small segment of the population for DNA sampling after a crime has occurred. Some fear that those chosen will often be minorities. A disturbing example occurred during an investigation for a serial rapist near Ann Arbor, Michigan, who was only described by the victim as a six-foot tall "light skinned, black man." Black men, who did not appear to be linked to the rapes through any evidence, were asked to provide blood samples for DNA analysis. If they refused, police obtained warrants to seize a sample. Although the police defended their "dragnet" procedure as necessary because there was no other evidence from which to identify a suspect, such discriminatory procedures can undermine public confidence in the criminal justice system and further stress already tense relationships between law enforcement and minority populations. A universal DNA database containing a DNA fingerprint from every citizen of the United States could be used to identify otherwise missed first-time offenders and to render unnecessary discriminatory dragnets. For universal DNA database searches to effectively identify as many people as possible, DNA samples from citizens would have to be obtainable regardless of an individual's consent. This would ensure the database would contain the most inclusive set of DNA fingerprints possible. There is some question regarding when during a person's life a sample would be obtained. One possibility is to require a sample when a person applies for a driver's license. It has also been suggested that samples could be taken from newborns while they are still in the hospital, similar to the blood spot or "Guthrie" cards that are currently created. Samples could also be taken as part of the routine vaccination requirements for children entering elementary school or couples applying for marriage licenses. Because there are many juveniles who commit crimes, a database consisting of samples taken from people at a very young age would be the most effective for crime detection. Obtaining samples from young people would also enhance the ability of the universal database to be used for identifying young disaster victims.

## DNA Databases and the Fourth Amendment

In general, DNA databases have withstood constitutional challenge on a number of fronts. The most difficult constitutional issues concerning databases are Fourth Amendment challenges. While existing offender databases have withstood attacks under the Fourth Amendment, a universal database would not withstand a Fourth Amendment challenge under existing Supreme Court precedent. To understand why creating a universal database would be unconstitutional, it is necessary to examine both the Fourth Amendment generally and how courts have applied it offender DNA databases. Although the Fourth Amendment does not protect people from all searches and seizures, it does require that searches and seizures be reasonable. Generally, for a search or seizure to be reasonable, it must be conducted under the authority of a warrant issued by a neutral magistrate upon a showing of probable cause. As with most constitutional doctrines, the contours of the Fourth Amendment's warrant requirement have evolved, resulting in exceptions to the general rule. Searches and seizures have been upheld as reasonable in situations where probable cause existed but obtaining a warrant was impractical; when individualized suspicion existed but both probable cause and a warrant were missing; and when all three, a warrant, probable cause, and individualized suspicion were missing but a "special need" that extended beyond normal law enforcement purposes was present.

Courts have upheld statutes authorizing forcible extraction of DNA from offenders for use in offender DNA databases without a search warrant, probable cause, or individualized suspicion through two separate lines of reasoning, both beginning with the assumption that drawing blood or extracting a tissue sample constitutes a search and seizure.

## A. The Balancing Approach

One of the first cases to examine a statute authorizing the creation of an offender DNA database under the Fourth Amendment was Jones v. Murray. In Jones, six inmates under the custody of the Virginia Department of Corrections challenged that state's statute creating an offender DNA database and its procedure for extracting samples. The inmates' primary argument was that the statute, by authorizing a search and seizure of their bodily fluid without individualized suspicion that they had committed a crime, constituted an unreasonable search and seizure and therefore violated the Fourth Amendment.

The Fourth Circuit agreed that the extraction of a blood sample was a search under the Fourth Amendment and therefore would only be justified if it was reasonable. The court, however, rejected the inmates' argument that a search could not be reasonable if conducted without individualized suspicion. The court noted that there was no "per se Fourth Amendment requirement of probable cause, or even a lesser degree of individualized suspicion, when government officials conduct a limited search for the purpose of ascertaining and recording the identity of a person who is lawfully confined to prison." The court then distinguished "free persons" from "confined persons," noting that in the case of confined persons probable cause that a crime had been committed was already established, as evidenced by an arrest or a conviction. Thus, the offender had a lessened expectation of privacy during his confinement. The court also noted that once a suspect has been arrested, the state develops a legitimate interest in his identity and connection to the crime at issue as well as a means of solving future crimes. Because of this interest, it is standard booking procedure to take fingerprints even if such prints will not be used to link the suspect at trial to the crime for which he was arrested. Since fingerprints are collected without individualized suspicion to solve future crimes, and the court determined that taking a blood sample to create a DNA fingerprint was sufficiently similar to a traditional fingerprint, the court reasoned there was no need for a finding of individualized suspicion for the blood test to be reasonable. The court then considered the extent of intrusion into the prisoner's lessened expectation of privacy. Although recognizing that the invasiveness of the procedures to obtain standard fingerprints and DNA fingerprints were vastly different, the court nonetheless found the blood tests to be "minimal intrusions" of privacy and bodily integrity. The court then examined the purported government interest in obtaining DNA samples for future law enforcement purposes. The court noted that DNA provides a "dramatic new tool for the law enforcement effort to match suspects with criminal conduct" by providing an unalterable means of similar but more accurate identification than basic fingerprinting. Finding the detection and prevention of future crimes through DNA fingerprinting to be a strong government interest, the court ultimately determined that the government interest outweighed a prisoner's lessened expectation of privacy and the minimal intrusion of obtaining the blood sample. In addition to the Fourth Circuit, other courts that have followed this balancing approach have found that an individual's Fourth Amendment interests in privacy are lessened when in police custody such that they are outweighed by the more significant government interests in the detection, identification, and prevention of crime. Although offender DNA databases are constitutional under this approach, the establishment of a mandatory universal DNA database containing samples obtained from "free people" probably would not survive a constitutional challenge. The privacy rights of the newborn or the driver's license applicant coupled with the intrusion of the sampling procedure would be weighed against the government interest in investigating and preventing future crimes. Because both the newborn and the drivers' license applicant are "free persons" and not confined, they would both have greater expectations of privacy than would a prisoner. Even when analyzed with a less intrusive sampling procedure that collects saliva in lieu of blood, dicta in prior Fourth Amendment cases discussed below suggests that a court considering the issue would find the individual privacy interests more compelling and determine these searches to be unreasonable.

In Davis v. Mississippi, the Supreme Court noted that mass "dragnet" fingerprinting campaigns without warrants, probable cause, or individualized suspicion are unreasonable searches and would violate the Fourth Amendment. Although dicta in Davis has been interpreted not to require full probable cause for certain limited investigatory seizures and subsequent fingerprinting, Davis has been read to require at least some form of suspicion before the seizure and fingerprinting is conducted. Although one commentator has suggested that DNA samples could be obtained from individuals for DNA "line-ups" based on a lesser reasonable suspicion standard, even this suggestion requires some quantum of suspicion that the individual being sampled has committed a crime. Such a suspicion of involvement in criminal misconduct would be missing for most citizens when law enforcement collected DNA fingerprints from every individual for inclusion in a universal DNA database. Further, the Supreme Court acknowledged in Schmerber v. California, that "the interests in human dignity and privacy which the Fourth Amendment protects forbids any such intrusions on the mere chance that desired evidence might be obtained." The language in these cases suggests that, although courts are willing to allow minimal intrusions into the privacy of confined individuals – i.e., those who have effectively abdicated most of their privacy expectations through the commission of a crime courts are less tolerant of mass searches conducted without any quantum of belief that an individual has been involved in a crime, regardless of the extent of the privacy intrusion.

## B. The "Special Needs" Approach

As previously noted, the general requirement that a search be conducted only after a warrant supported by probable cause has been issued, can be eliminated in certain limited circumstances. The Supreme Court in New Jersey v. T.L.O. established a "special needs exception" to the warrant and probable cause requirements. The exception, as restated in subsequent cases, is available "where a Fourth Amendment intrusion serves special governmental needs, beyond the normal need for law enforcement." Like the balancing test, the special needs exception requires the court to consider the individual's privacy expectations and the government's interests.

Unlike the basic balancing approach, however, the special needs exception does not require a warrant, probable cause, or individualized suspicion if, in the particular context, it would be impractical or would impede the purpose of the search.

In State v. Olivas, the Washington Supreme Court, considering a Fourth Amendment challenge to a statute authorizing the extraction of blood samples to created an offender DNA database, used the special needs exception to uphold the statute. The court began by recognizing that drawing blood for the sample is a search and seizure under both Schmerber and Skinner. The court then noted that two studies in Virginia had shown that approximately 62.5% of all felons were arrested for an additional felony or serious misdemeanor within three years of their release from prison, and that recoverable DNA is left at the scene of a violent felony approximately thirty percent of the time. Prisoners' knowledge that their DNA fingerprints are in a database, making them easily identifiable, could deter those same prisoners from re-offending after their release. Without further discussion, the court concluded that the government interest in deterring recidivist criminal activity through the use of an offender DNA database was a need "beyond normal law enforcement" and therefore qualified as a "special need."

A universal database, under the reasoning of Olivas, would not be defendable under the special needs exception. The case law suggests that the special needs exception was meant to apply to the limited set of circumstances where the purposes of requiring a warrant and probable cause would not be furthered, and where the privacy rights of the individual are not unduly infringed in light of the government interest.

Special needs are defined as those needs of the government that do not occur in the normal course of law enforcement. The majority in Olivas appears to have adopted the idea that deterring recidivism and identifying and prosecuting future offenders through the use of offender DNA databases is acceptable does not fall within the normal course of law enforcement; this court reached this conclusion, however, without discussion and over the objection of Justice Utter's concurrence. In light of the growing movement toward community policing, the argument that deterring crime is not a regular duty of law enforcement is becoming increasingly difficult, if not impossible, to make.

Allowing searches to assemble a universal DNA database would ignore one of the critical bases for the Supreme Court's decisions, the lessened expectations of privacy that flows from an individual's employment or status as an offender. The Supreme Court has ruled that neither individuals participating in a highly regulated industry, nor minors in a public school who participate in extracurricular activities expect their privacy rights to be fully protected because these individuals voluntarily took some action that allowed the government's interest in safety or drug abuse detection to outweigh the their (already reduced) expectation of privacy. Searches conducted for the sole purpose of collecting samples for inclusion in a universal DNA database would lack this initial action by the individual whose privacy interest and expectation is otherwise still intact.

## Universal DNA Databases and the Danger of the Preventive State

While allowing the government to maintain a database of DNA fingerprints from every citizen could lead to more crimes being solved and, potentially, even more being deterred, we must consider at what price?

> *The ultimate question, plainly, is a value judgement Whether, if the particular form of surveillance practiced by the [government] is permitted to go unregulated by constitutional restraints, the amount of privacy and freedom remaining to citizens would be diminished to a compass inconsistent with the aims of a free and open society.*

This Note argues that allowing the government to effectively search and seize the DNA of every citizen without a warrant, probable cause, or individualized suspicion that they committed a specific crime would damage how we perceive ourselves as a free and democratic society. As a democracy, our government gains its power and legitimacy from its relationship with its citizens. This delicate balance is best described as a relationship of trust between the government and its citizens: the government trusts its citizens to act within the law, and the citizens trust the government to allow them to go about their lives protected by an appropriate zone of privacy into which the government does not interfere. Scott Sundby argues that this trust relationship is hindered when "the government is allowed to intrude into the citizenry's lives without a finding that the citizenry has forfeited society's trust to exercise its freedoms responsibly."

The Fourth Amendment is an example of how the Framers intended for this trust relationship to be maintained. The Fourth Amendment was formulated in response to the experience of the Framers as colonists prior to the Revolution with general warrants, known as "writs of assistance." These general warrants granted customs officials expansive authority to search homes, places of business, and vessels for potential violations of the British tax laws, without any particularized belief that contraband would be there. Such general warrants were vehemently denounced by James Otis in 1761 as, "the worst instrument of arbitrary power, [and] the most destructive of English liberty." The Supreme Court has also expressed mistrust of arbitrary power. In 1921, Justice Clarke argued that the Fourth Amendment is "indispensable to the full enjoyment of personal security, personal liberty, and private property; to be regarded as of the very essence of constitutional liberty." Justice Brandeis articulated that "the right to be let alone," the value protected by the Fourth Amendment "is the most comprehensive of rights and the right most valued by civilized men."

By requiring the government to obtain a warrant supported by probable cause, the Fourth Amendment increases citizen control over when the government can intrude into his privacy. Government interference into people's privacy at the whim of the state is a characteristic of a totalitarian society. One scholar has described a police state as a society in which the "ominous rap on the door at night" is typical. As Justice Jackson observed: "Uncontrolled search and seizure is one of the first and most effective weapons in the arsenal of every arbitrary government. The human personality deteriorates and dignity and self-reliance disappear where homes, persons and possessions are subject at any hour to unheralded search and seizure by police." The Fourth Amendment is meant to protect against this type of intrusion by limiting when the government may interfere in people's privacy, and the extent to which the government may intrude. The freedom from these unconstitutional intrusions allows all other freedoms to be exercised more extensively. These freedoms allow Americans to freely engage in political debate, and consent to government actions, activities that embody the essence of our democratic society.

Allowing the government to maintain a database of every individual's genetic information creates, at the very least, a government that knows more about its citizens. Even if the information is limited to identifying fingerprints, genetic or traditional, without more intrusion into the private thoughts and minds of individuals, there is still something ominous and oppressive about an all-knowing state. The inherent danger to our conception of ourselves as a free and autonomous society requires that further expansion of the preventive state, represented by the creation of a universal database, be vigorously opposed.

## Conclusion

The use of DNA fingerprinting as a forensic tool has undergone a dramatic change. While initially limited to confirming the identity of an individual already suspected of committing a specific crime, the use of offender DNA databases has altered the way a criminal investigation can proceed. Minuscule amounts of DNA recovered from a crime scene can be used to link an otherwise unknown suspect to the crime. While existing offender DNA databases have been upheld over Fourth Amendment challenges, they have survived largely because of the minimal privacy expectations offenders have due to their status as offenders. Extending this principle to the context of free persons who are not suspected of any crime is unwarranted. Attempts to implement a universal DNA database containing DNA fingerprints from every member of society would fail to satisfy constitutional scrutiny because free persons have no diminished expectation of privacy.

Allowing a universal DNA database would eviscerate the delicate trust relationship between American citizens and their government. Allowing the state to intrude without suspicion that the individual violated a law would allow the government to intrude capriciously into citizen's zones of privacy.

# 8

# Forensic DNA Kits

## Introduction

When forensic DNA testing arrived in the courtroom, it was heralded as the "greatest advance in crime fighting technology since fingerprints." From the beginning, however, non forensic scientists and defense lawyers expressed concern that enthusiasm for the new technology had outpaced forensic laboratory capabilities. This concern arose from the fact that many labs were suffering from chronically high error rates even before crime laboratories were asked to perform the highly sophisticated analysis forensic DNA testing requires. To find out if labs suffer from similar error rates while performing forensic DNA analysis, defendants have sought extensive discovery of testing data. The laboratories performing DNA testing have resisted discovery.

The discovery conflicts surrounding forensic DNA testing moved into a new arena when state laboratories began using DNA kits prefabricated by private corporations. Commercial forensic DNA kits are like chemistry sets: they include all of the materials used in DNA testing and detailed instructions on how to produce results. The kits provide a protocol and ingredients, but the companies are generally reluctant to reveal information unnecessary to the step-by-step testing process. In fact, the companies provide so little information that analysts using kits have been compared to "the average person using a VCR – with no knowledge of the science that makes it work, but happy to push a button and get results."

Thus, when defendants seek to discover information about a kit, they must subpoena it from the private corporation that makes the kit. Currently, the primary manufacturers of the kits are Promega and Perkin-Elmer Corporation , successors in interest (at least partly) to DNA typing products originally developed by the Cellmark and Cetus corporations, respectively. Like forensic laboratories, PE and Promega initially resisted defense efforts to obtain discovery of data. Promega, however, eventually published the data that defendants sought." PE, while making limited concessions in individual cases, has continued to fight discovery in other cases.

In fighting discovery, both PE and Promega have advanced (and PE continues to advance) various arguments. They have relied most heavily on the argument that the information sought is proprietary and thus protected under the trade secret privilege. Alternatively, they have contended that defense subpoenas of data are invalid because the burden the corporations would suffer in gathering requested information outweighs the defendant's need for the information.

## Challenging the Reliability of Forensic DNA Kits

Criminal defendants have sought access to developmental validation data and primer sequences from DNA kit makers to effectively challenge the results of DNA testing in court. The discussion in this Part begins by describing in general terms the two ways in which a defendant can challenge the reliability of scientific evidence: challenging its admissibility at a pretrial admissibility hearing and challenging its weight before the jury at trial. The discussion then moves to specifics, describing the types of data that defendants are seeking from kit makers and how that evidence would assist defendants in making those challenges.

### A. Admissibility

Under the law of evidence, trial courts must evaluate the admissibility of scientific evidence. Although the required threshold showing differs among jurisdictions at least formally, the underlying concern of courts in all jurisdictions in evaluating the admissibility of evidence is trustworthiness, also called "evidentiary reliability." There are two predominant tests used to determining the reliability of scientific evidence: the Daubert test and the Frye test.

1. Daubert. In the 1993 case Daubert v. Merrell Dow Pharmaceuticals, Inc., the Supreme Court held that the Federal Rules of Evidence superseded the traditional standard for determining the admissibility of scientific evidence. Under Rule 702, scientific evidence is admissible only if it is shown to be reliable. The Court explained that evidence is reliable only where it is the product of sound scientific methodology and announced a nonexclusive list of factors by which a particular methodology may be evaluated. Those factors are: whether the underlying theory or technique can be and has been tested, whether the underlying theory or technique has been subjected to peer review and publication, the underlying theory's known or potential rate of error, and the general acceptance of the underlying theory in the scientific community. Thus, in Daubert jurisdictions, trial judges clearly are expected to determine for themselves the reliability of evidence and exclude evidence found to be unreliable.

At least one state trial court has held that DNA test results produced by Profiler Plus and PowerPlex kits are inadmissible on grounds of unreliability under Daubert. The Pfenning court in Vermont worried specifically about the fact that the manufacturers of the DNA kits had failed to disclose to both the defense and the scientific community at large the developmental validation studies and the primer sequences used to amplify the DNA before typing it. In discussing the refusal of kit makers to reveal information about the kits, the court wrote:

> *The failure of the manufacturers of DNA testing systems to disclose the primer sequences they have created to permit amplification of DNA is problematic from the perspective of scientific knowledge and, consequently, validation. It is more than problematic, it is anti-scientific in that it inhibits the ability of scientists in the field (including defense experts) to test the manufacturers' claims. Although the Court understands that the manufacturers believe they need to maintain as confidential what they consider to be proprietary information, in the case of new technology, it delays acceptance by the courts.*

Other courts in Daubert jurisdictions have not agreed with the above analysis and have instead held the results of DNA tests kits admissible despite the kit makers' refusal to publish certain data. For example, the Colorado Supreme Court recently held that results produced by kits ought not even be subject to Daubert admissibility hearings because they are merely applications of technology that has already been accepted. However, as the Vermont decision makes clear, the issue of the admissibility of the test results is not yet settled.
2. Frye. Until the Supreme Court's decision in Daubert, the standard used by federal courts and most state courts in determining the admissibility of scientific evidence was the Frye test. Although some states have chosen to adopt the new federal standard announced in Daubert, many states, including California, Florida, Michigan, and New York, have reaffirmed their allegiance to Frye. The District of Columbia Circuit Court of Appeals first articulated the Frye standard: "While courts will go a long way in admitting expert testimony deduced from a well-recognized scientific principle or discovery, the thing from which the deduction is made must be sufficiently established to have gained general acceptance in the particular field in which it belongs." The theory behind Frye is that general acceptance of scientific methodology serves as a surrogate for its reliability.

Although Frye could be read to require that trial courts merely confirm that a majority of the relevant scientific community's members supports the underlying principle at issue, many jurisdictions require a more substantive determination of reliability. For example, in Florida trial courts must consider the "quality, as well as quantity, of the evidence supporting or opposing a new scientific technique." To determine the quality of evidence, judges must make at least a limited inquiry into its reliability. Furthermore, Daubert's stricter standards have caused many Frye jurisdictions to take the trial court's gatekeeper role more seriously, resulting in greater attention to indicia of reliability traditionally outside Frye's scope of inquiry.

Like trial courts in Daubert jurisdictions, trial courts in Frye jurisdictions need not always hold an evidentiary hearing on the admissibility of scientific evidence. Frye courts need do so only where the scientific methodology underlying proffered evidence is novel.

Several courts have held that the evidence produced by DNA kits is not novel because kits are merely tools for carrying out accepted methodology. For example, in People v. Hill, a California appellate court rejected the argument that each new test kit must, as a matter of law, be subjected to a Frye hearing to determine its scientific reliability. Nonetheless, in most Frye jurisdictions courts have not yet decided whether particular models of DNA kits require Frye hearings, so trial courts are conducting hearings to avoid reversal. For example, in Florida, the trial court in State v. Yisrael conducted a Frye hearing to determine the reliability of the Profiler Plus and Cofiler Kits. Similarly, in Michigan, the trial court in People v. Phillips held that a Frye hearing was required because "new techniques applying generally accepted principles must themselves come to be generally accepted before their results may be received into evidence."

Although it is unclear if any jurisdictions will end up requiring Frye hearings for particular models of DNA kits, it is clear that trial courts in some jurisdictions are conducting such hearings. Furthermore, trial courts retain broad "latitude in deciding [how] to test an expert's reliability, and to decide whether or when special briefing or other proceedings are needed to investigate reliability." Where trial courts choose to hold Frye hearings on particular models of DNA kits, evidence of the kits' reliability will determine their admissibility.

## B. Weight

Even if courts decide not to hold hearings as to the admissibility of DNA kit results, defendants still will have the opportunity to challenge the reliability of the evidence before the jury at trial. All evidence related to reliability could be used by the defendant to challenge the weight given to the evidence by the jury, so long as the evidence is admissible. In holding that no Daubert hearing was required for particular models of DNA kits, the Colorado Supreme Court noted that all evidence bearing on the kits' reliability should "go to the weight of the evidence, rather than its admissibility."

## C. Data Underlying Forensic DNA Kits and Kit Reliability

Before addressing the ways in which the reliability of the kits is in question, it is important to note that many of the principles and methodologies underlying forensic DNA testing are no longer in dispute. Most importantly, all courts now accept the proposition that the structure of an individual's DNA is both constant and unique and thus may be used to match biological evidence at a crime scene with an individual. Courts also have accepted certain specific methods of typing DNA, that is, characterizing a segment of DNA according to either its sequence or length. DNA typing allows samples of DNA to be compared, and thus allows analysts to determine that a suspect and a biological sample share the same DNA profile.

The dominant methods of DNA typing now rely on Polymerase Chain Reaction, or PCR, which has been almost unanimously approved as an admissible methodology. PCR is an extraordinarily powerful means of replicating a chosen segment of a DNA sample, thus allowing more sophisticated tests to be performed. First, the technician extracts DNA from cells, and, using heat instead of enzymes, divides the strands from each other. The technician then adds a primer, a short piece of DNA, to the single strands. The primer "marks" the section on the DNA molecule that will be amplified. The enzyme polymerase is added to the primer-DNA mixture. The polymerase, like naturally occurring enzymes that result in the DNA replication required to create new cells, creates a new complementary strand to bond with each of the original strands. The process is repeated until large amounts of DNA are created. After PCR, the DNA may be analyzed with both sequence-based methods and length-based methods. Furthermore, PCR-STR, the method used by PowerPlex, Profiler Plus, and Cofiler, and promoted by the FBI as part of its Standardization Project, is also being rapidly accepted by courts.

Even though courts have accepted the basic science underlying forensic DNA testing, not all of them have accepted that commercially produced forensic DNA testing kits are reliable. To effectively challenge kits' reliability, defendants need the following types of data.

1. Developmental Validation Studies. Developmental validation determines the limitations of a new technology, such as what kinds of errors a technology is prone to and what types of precautions must be taken to ensure that errors do not occur. Defendants seek

access to data produced in developmental validation studies to ensure, on the most basic level,' that developmental validation was in fact performed. It is uncontroverted that developmental validation determines whether a given method or kit produces accurate, reliable results. As the first National Research Council report on forensic DNA technology noted, "there is no scientific dispute about the validity of the general principles underlying DNA typing. However, a given DNA typing method might or might not be scientifically appropriate for forensic use." As a result, the report recommends that any new DNA typing method, or variation on an existing method, be rigorously tested to determine its reliability. The report's recommendation has been seconded by both the Technical Working Group for DNA Analysis Methods (TWGDAM) and the DNA Advisory Board . Furthermore, the National Institute of Standards and Technology has addressed STR systems like Profiler Plus, Cofiler, and PowerPlex specifically: "Before a new STR system or STR multiplex may be routinely employed in human identity testing it should be extensively validated to insure reliability of results." Thus, discovery could reveal that the required validation has not in fact been performed.

Even if discovery shows that the kits were appropriately developmentally validated, the data produced in that process could reveal weaknesses of a particular kit model. For example, specific DNA tests are associated with specific artifacts. Artifacts are byproducts of the testing process that could be mistaken for evidence of a specific genetic makeup. During developmental validation, the kit should be "rigorously characterized with respect to the types of possible artifacts, the conditions under which they are likely to occur, the scientific controls for detecting their occurrence, and the steps to be taken when they occur." The data produced by this study would thus help the defense recognize artifacts that the analyst may have misinterpreted, know whether the laboratory used the proper controls to detect the occurrence of artifacts, and understand whether the analyst responded properly to signs that an artifact may have occurred.

The developmental validation data also would reveal the kit's ability to discern the results of environmental influences on the sample. According to TWGDAM guideline 4.1.5.6, developmental validation studies must include environmental studies, "so that the effects of factors such as matrix, age, and degradative environment (temperature, humidity, UV) on a sample are considered." The data produced by the developmental validation studies would reveal the kit's ability to distinguish between true results of the DNA analysis and results that reflect environmental degradation.

2. Primer Sequences. The primer sequences are as material as developmental validation to the defense's case that the kits are unreliable. Primers are used in PCR, the amplification step of DNA testing. DAB guideline 4.1.4 requires that "the DNA primers selected for use in the forensic DNA analysis must be readily available to the scientific community." This is at least partly because the extraordinary sensitivity of the PCR process makes it highly vulnerable to contamination. As the National Research Council has made clear, "any procedure that uses PCR is susceptible to error caused by contamination leading to amplification of the wrong DNA." Contamination may occur from several sources, including handling in the field during collection, cross-contamination of samples in the laboratory, and carryover contamination (the contamination of evidence samples or reaction solutions with PCR products from prior amplifications). Most relevant to the discovery rights at issue here, contamination may be caused by the kits themselves. Because of the sensitivity of the PCR process, the primers, which contain DNA, could contaminate the results. Without knowing the sequence of the primers, neither defense experts nor courts could determine if the effects of contamination by the primer sequence are present in the results.

Lastly, knowledge of the primer sequences would help to explain a specific problem with the Profiler Plus kit that has arisen. In one study of the vWA locus, PE's Profiler Plus Kit and the Promega PowerPlex Kit produced inconsistent genotyping. To understand why the tests are producing different results, PE is sequencing the locus. Without access to the primer sequence used in the kit, defense experts are unable to perform similar testing.

## The Current Discovery Regime

There is currently no nationally recognized right of access to the data underlying DNA kits. In the federal system and in the vast majority of states, discovery is controlled by statutes or court rules. Although the statutory schemes are similar in structure, they vary greatly in content. As a result, the status of discovery rights varies among jurisdictions. Most of these rules are quite restrictive of discovery in the criminal area. Discovery in criminal prosecutions traditionally has been extremely limited and remains far less robust than discovery in civil proceedings.

There are a few common provisions that could be read to support discovery of the data at issue here, but they have been applied in a manner that limits their usefulness. Federal Rule of Criminal Procedure 16 and almost all state discovery provisions mandate the disclosure of scientific reports. Although the phrase "scientific report" could be interpreted to include any information relied upon, either explicitly or implicitly, in creating the final report, most jurisdictions have not interpreted the rule in this way. In fact, some states have specifically shielded from discovery "underlying documentation" used in preparing a final report. For example, the Ninth Circuit has made clear that log notes produced in the testing process are not discoverable as scientific reports. Thus, Rule 16 and its counterparts present an unsure avenue by which to pursue discovery of this evidence.

The other discovery rule that could be read to allow for discovery of the developmental validation data and primers is Federal Rule of Criminal Procedure 16. Rule 16 and the rules patterned after it allow discovery of documents and objects "which are material to the preparation of the defendant's defense." In jurisdictions in which documents are not discoverable as scientific reports, courts have held that they still may be discoverable under Rule 16. For example, in United States v. Yee, the court, relying on Rule 16, held discoverable laboratory matching criteria and standards, environmental insult tests, population data, and proficiency testing data held by the FBI. However, the application of Rule 16 is limited to those things within the government's control. It does not cover evidence held by third parties. Thus, as long as the kit makers decline to share their information with the crime labs using their kits, Rule 16 may not grant the defense discovery.

There is currently no reliable scheme in place that provides for the discovery of data underlying DNA kits. However, denying a defendant access to this data may violate the Constitution. If the underlying scientific data is not provided, and the evidence provided by the kits is admitted, the defendant's rights under the Compulsory Process, Confrontation, and Due Process Clauses are violated.

## Arguments Against Expanded Criminal Discovery

### A. The Traditional Arguments Against Criminal Discovery

The traditional arguments against criminal discovery are premised on the idea that the Bill of Rights's protections for the accused skew the adversarial system in the defendant's favor. The most succinct expression of this argument was made by Judge Hand:

> *Under our criminal procedure the accused has every advantage. While the prosecution is held rigidly to the charge, he need not disclose the barest outline of his defense. He is immune from question or comment on his silence; he cannot be convicted when there is the least fair doubt in the minds of any one of the twelve. Why in addition he should in advance have the whole evidence against him to pick over at his leisure, and make his defense, fairly or foully, I have never been able to see.*

Hand's first contention, that the defendant need not disclose "the barest outline of his defense," has become less compelling as states have moved toward allowing prosecutorial discovery. The expansion of prosecutorial discovery gained momentum in the 1970s and currently a majority of states have discovery rules that provide the same rights of discovery to the prosecution as to the defense. Hand's second argument against criminal discovery relates to the defendant's right to remain silent. Hand obviously considered the Fifth Amendment a significant advantage to the defendant in the adversarial process. However, as numerous commentators have pointed out, police officers regularly obtain incriminating statements from criminal defendants during pretrial interrogations, despite Miranda warnings. Another commonly raised argument against liberal criminal discovery rests on the fear that criminal defendants will attempt either to intimidate witnesses or obtain perjured testimony that contradicts the prosecution's case. Regardless of this argument's merit in other contexts, its merit in the context of scientific evidence is certainly limited. The significance of witness intimidation is reduced where a witness's testimony is reproducible. If one expert backs out, the evidence could be tested anew by a braver soul. The possibility for perjury as to the content of scientific evidence is equally limited; as the ABA has noted, "it is virtually impossible for evidence or information of this kind to be distorted or misused because of its advance disclosure."

## B. The Trade Secret/Innovation Argument

A more compelling argument against liberal discovery of the scientific basis of DNA kit testing rests on the possibility that forced disclosure could result in less, or slower, innovation in forensic sciences in the future. The kit makers have implicitly relied on such a rationale in arguing that the data sought is protected from discovery as a trade secret. The trade secret privilege, like all intellectual property law, is based in part on the premise that protection of property is required to foster innovation. The trade secret privilege has been upheld in the context of DNA kit testing by a trial court in Arizona, among others.

The kit makers are certainly correct in emphasizing the role that private innovation played in the development of forensic DNA technology. Lifecodes and Cellmark Laboratories, both private corporations, began forensic testing in 1987, a year before the FBI opened the first publicly operated DNA typing lab. The first DNA evidence introduced in court was the result of a Lifecodes Corporation test, and the first death penalty conviction involving DNA evidence was achieved using DNA evidence produced by Cellmark Diagnostics. The FBI's decision to open a DNA laboratory was based in part upon the desire to reduce the role of corporations in forensic testing. Despite the opening of public laboratories, private corporations remained central to innovation in forensic DNA technology. The opening of public laboratories ensured that most actual analysis would be conducted by public employees, but the equipment and kits used by state laboratories continue to be produced by corporations.

Nevertheless, the appropriateness of applying the trade secret privilege to kit makers is complicated by the fact that the privilege is far more developed in civil litigation than it is in criminal law. The discrepancy is best explained by the history of trade secret misappropriation law, which provides the forum in which parties are most likely to claim a trade secret privilege. Until recently, trade secret law provided a remedy for trade secret misappropriation only through civil litigation.

In contrast, the boundaries of the trade secret privilege in criminal cases remain less clear, largely because the privilege rarely arises in the criminal context. Although more than half of the states now protect trade secrets through criminal statutes, state prosecutors rarely use them. Nor has the Economic Espionage Act, the federal criminal trade secret law, produced a barrage of cases; prosecutions under the Act have been few because Attorney General Janet Reno required that charges under the Act not be brought without her personal approval or the approval of her deputy until October 2001. The Federal Rules of Criminal Procedure also provide limited guidance, allowing only that: "Upon a sufficient showing the court may at any time order that the discovery or inspection be denied, restricted, or deferred, or make such other order as is appropriate."

The role of the trade secret privilege in criminal cases has been addressed by at least one state legislature. Under California evidence law, the trade secret privilege does apply to criminal cases. However, the privilege is not absolute: "No evidence, however, shall be excluded during a criminal proceeding pursuant to this section if it would conceal a fraud, work an injustice, or deprive the People or the defendant of a fair trial."

Assuming for the moment that PE is correct in asserting that the developmental data and primer sequences are trade secrets, a balancing of equities makes clear that the data should be revealed at least to the defendant and his expert witnesses. Here, the denial of access to the claimed trade secrets is highly prejudicial; in fact, it severely infringes on the defendant's constitutional rights. Moreover, although it may be true that the dissemination of the information could damage kit makers economically, it also must be noted that kit makers are under a scientific obligation to release this information for peer review. The TWGDAM guidelines state: "It is essential that the results of the developmental validation studies be shared as soon as possible with the scientific community through presentations at scientific/professional meetings. It is imperative that details of these studies be available for peer review through timely publications in scientific journals."

Lastly, protective orders are available to limit public access to the information. Protective orders are, of course, unsatisfying to both parties. PE and other kit makers fear that they will be violated. Defense lawyers argue that secrecy impedes understanding of the tests and contravenes scientific principles of peer review. As one commentator put it, "If scientific evidence is not yet ready for both scientific scrutiny and public re-evaluation by others, it is not yet ready for court."

In the end, the question becomes: Do we want to encourage technological innovation at the expense of the constitutional rights of defendants? As importantly, does forensic technology increase the ability of the system to find truth where the evidence cannot be effectively challenged by the defense? Even the most discerning forensic technology will sometimes produce inaccurate results.

## C. The Argument that the Burden of Collecting the Information Is Too Great

At least one trial court, People v. Bertsch & Hronis, has allowed the prosecution to use PE kit results despite PE's refusal to disclose the underlying data because it believed that compliance with the defendant's subpoena would be unduly burdensome. The court agreed with PE that "the burden on the third party ... outweighs substantially the alleged demonstrated need as set forth." The court felt that the defendant's need was not substantial, and certainly did not rise to the level of a constitutional defect, even though the defendant had no access to the primer sequences and only partial access to developmental data. The court did not indicate how much developmental data was provided.

Not only did the court underestimate the burden that lack of access would impose on the defendant, it also overestimated the burden that production of the data would create for PE. The court said: "it is a significant burden involving several people, four people over several weeks." While this may be true, PE should have anticipated such a request and better organized its data. Even though PE has not yet published its developmental data, the PE "Users Manual" that it distributes with its kits promises peer review publication of the developmental data. Furthermore, both TWGDAM and DAB Guidelines call for publication of developmental validation. Practically speaking, one must question the burden on an enormously wealthy company like PE of hiring a few people for a few weeks to collect and organize the data.

## D. The Withdrawal from the Market Argument

In Michigan, where a panel of three district court judges recently considered the admissibility of Profiler Plus test results, law enforcement openly worried that a discovery order would lead PE and other biotechnology companies to withdraw their products from the forensic science market. Such fears are reinforced by the fact that forensic testing kits and equipment account for only ten percent of the sales of Applied Biosystems (the division of PE that markets forensic DNA kits).

There are indications, however, that law enforcement would remain a profitable market for PE. Although revelation of the primer sequences and developmental validation data creates the opportunity for labs to use PE technology without purchasing a kit, many labs will continue to use kits because of the kits' efficiency. It is possible to create one or two primers by hand, but forensic scientists at overburdened labs appreciate the shortcut that pre-prepared primers provide.

Even if the forensic market remains profitable, corporations may prefer to abandon that market if revelation of the data would severely harm their competitiveness in other life science industries. In the end, however, the best indication of an industry's ability to withstand revelation of proprietary information is provided by the industry itself. Promega, the other provider of STR multiplex kits, already has indicated that it is willing to sacrifice secrecy for access to the forensic market. If a particular corporation feels it cannot afford to reveal the information, it seems highly likely that another will emerge to take its place.

## Constitutional Implications of the Lack of Discovery of Developmental Validation Data and Primer Sequences

The Supreme Court has noted that "there is no general constitutional right to discovery in a criminal case." Unfortunately for the clarity of the Court's holdings on discovery and constitutionality, the statement fails to accurately describe the Court's own holdings under the Compulsory Process, Confrontation, and Due Process Clauses; and undercuts the values embodied in these three clauses, particularly as those values relate to the reliability of scientific evidence. Not only is there a constitutional right to discovery, that constitutional right applies to the data that supposedly demonstrates the scientific reliability of DNA evidence used against defendants. The Constitution requires that test results not be admitted unless defendants are given access to the scientific data alleged to establish the kits' reliability.

## A. Compulsory Process

To date, no trial or appellate court has considered whether admitting DNA evidence against a defendant who lacks access to the data underlying the DNA testing kit violates the Compulsory Process Clause. However, as the following discussion demonstrates, admitting DNA evidence in these circumstances violates the dictates of the Compulsory Process Clause and undercuts the values it represents.

The Compulsory Process Clause of the Sixth Amendment establishes that "in all criminal prosecutions, the accused shall enjoy the right ... to have compulsory process for obtaining witnesses in his favor." At a minimum, compulsory process means "that criminal defendants have the right to the government's assistance in compelling the attendance of favorable witnesses at trial and the right to put before a jury evidence that might influence the determination of guilt." There are two issues here. First, to what degree, if any, does the Compulsory Process Clause include a right of discovery? Second, if such a discovery right exists, is it applicable to developmental validation and primer sequences held by private corporations?

The two cases that deal most directly with the rights of discovery created by compulsory process are United States v. Nixon and United States v. Burr. In Nixon, the Court unanimously upheld an order requiring President Nixon to produce tapes of his personal conversations with the Watergate defendants. The defendants' right of discovery was grounded at least partly in the Compulsory Process

Clause: "The right to the production of all evidence at a criminal trial similarly has constitutional dimensions. The Sixth Amendment explicitly confers upon every defendant in a criminal trial the right ... 'to have compulsory process for obtaining witnesses in his favor.'"

The second significant case is United States v. Burr. Burr was on trial for treason and sought discovery of a letter from General Wilkinson to President Jefferson. Burr wanted the letter because President Jefferson claimed to rely upon it when he announced to Congress that he believed Burr to be guilty of treason. President Jefferson's main objection to the subpoena duces tecum was that it arrived prematurely, before the grand jury indicted Burr. In rejecting this argument, Marshall explicitly relied upon the Compulsory Process Clause as a source of pretrial discovery:

> *Upon immemorial usage, then, and upon what is deemed a sound construction of the constitution and law of the land, the court is of [the] opinion that any person charged with a crime in the courts of the United States has a right, before as well as after indictment, to the process of the court to compel the attendance of his witnesses.*

In holding the subpoena valid, Marshall also made clear that the word "witnesses" includes written materials. He rejected any distinction between people and things under the Compulsory Process Clause. The question remains whether the Compulsory Process Clause creates a right of discovery applicable to the data and developmental validation studies underlying DNA kits. In United States v. Nixon, the Court explained that parties in criminal trials who move for a subpoena duces tecum "must clear three hurdles: relevancy; admissibility; specificity."

Here, the information is undoubtedly relevant. The prosecution, through expert testimony, will want to show that the kits provide reliable and accurate information, and defendants will want to show, through the developmental validation study and primers, that the kits either are not generally reliable or failed to provide reliable information in a particular case.

Next, the movant must show that evidence of the developmental validity study and primers is admissible. This information is almost certainly admissible during a pretrial hearing to determine the admissibility of DNA evidence. Under Federal Rule of Evidence 104, a judge's admission of foundational or preliminary facts is not bound by rules of evidence except with respect to privileges. Courts have uniformly held this rule to override the general applicability to preliminary facts of exclusionary rules such as hearsay. The admissibility of validity studies at trial is not quite as clear-cut, but the great probative value of this evidence makes it likely that courts will find it admissible The most important use of this evidence may very well be during the pretrial hearing, where the evidence will be used to determine whether the jury ever even hears the results of scientific testing. Its admissibility before trial surely satisfies the requirements of Nixon, which interprets the Compulsory Process Clause to be concerned primarily with the effective functioning of the adversarial system.

The third requirement relates to the specificity of the request. The request for discovery of developmental validation and primer sequences is highly specific. Despite PE's protestations to the contrary, the data is both discrete and of a type regularly produced during a kit's validation.

Lastly, an understanding of the Compulsory Process Clause that mandates disclosure of the data makes sense in light of the policy goals identified in cases interpreting the clause. The Supreme Court has written that "the right to offer the testimony of witnesses, and to compel their attendance, if necessary, is in plain terms the right to present a defense ... to the jury so it may decide where the truth lies." Here the Court identified two interrelated themes in Compulsory Process Clause jurisprudence: adversariness and truth seeking. Increased access to the data underlying DNA kits serves both those goals. It enables defendants to challenge the evidence against them, and it enables the court to understand better the value of DNA evidence.

## B. Confrontation Clause

Like the Compulsory Process Clause, the Confrontation Clause has been noticeably absent from any discussion of the discoverability of kit data. Even though this absence is perhaps more understandable, because the Supreme Court recently has limited sharply the Clause's application to pretrial discovery, there is a plausible argument that even under the Court's narrowed understanding of the Confrontation Clause, the conflict at issue here constitutes an infringement of defendants' Confrontation Clause rights. Furthermore, the values embodied by the Confrontation Clause are as much at risk in denial of discovery as are the values embodied in the Compulsory Process Clause.

The Supreme Court has not been generous to criminal defendants in interpreting the mandates of the Confrontation Clause. Whereas the Compulsory Process Clause has been read to include pretrial discovery rights, the Court has described the Confrontation Clause as follows:

> *The opinions of this Court show that the right of confrontation is a trial right, designed to prevent improper restrictions on the types of questions that defense counsel may ask during cross examination. The ability to question adverse witnesses, however, does not include the power to require the pretrial disclosure of any and all information that might be useful in contradicting unfavorable testimony.*

In so holding, the Ritchie Court's plurality opinion made clear that the Confrontation Clause is "normally" satisfied so long as "defense counsel receives wide latitude at trial to question witnesses." The above formulation of the Confrontation Clause raises the question of which cases are "normal." As Justice Blackmun noted in his concurrence, the plurality's decision to divorce Confrontation Clause rights from discovery rights amounts to a dismissal of any inquiry into "the effectiveness of the cross-examination" and renders confrontation rights, at least in some cases, an "empty formality." The plurality's understanding of the Confrontation Clause creates the greatest obstacle to effective cross-examination of expert witnesses. The complexity of many types of expert testimony, such as testimony on DNA testing, makes pretrial preparation imperative to effective cross-examination. Forensic scientists agree. According to Douglas Lucas, director of the Canadian Centre for Forensic Sciences, "'If cross-examination is to be the only way to discover misleading or inadequate testimony by forensic scientists, then too much is being expected from it.'" The extreme difficulty of effectively cross-examining an expert witness without pretrial preparation and discovery raises the issue of whether cross-examination of experts raises "normal" Confrontation Clause issues within the meaning of the Ritchie Court's plurality opinion.

Also, refusing defendants access to the scientific bases of evidence used against them contravenes fundamental Confrontation

Clause values. The Confrontation Clause values with which the Supreme Court most consistently has concerned itself are the reliability of the evidence presented and the jury's ability to evaluate the strength of that evidence. These values are clearly set out in the Supreme Court's cases on hearsay. In Ohio v. Roberts the Court explained that the Confrontation Clause allows for the admission of an out-of-court statement by an unavailable declarant only if the statement contains "indicia of reliability." The Roberts Court also held that the Confrontation Clause is concerned with "`affording the trier of fact a satisfactory basis for evaluating the truth of the prior statement.'" The Court summed up its concern with both the reliability of the evidence presented and the jury's ability to evaluate that evidence in Greene v. McElroy: "The evidence used to prove the Government's case must be disclosed to the individual so that he has an opportunity to show that it is untrue."

Would the values described above be served by allowing defendants discovery of data underlying DNA kit results? Access to developmental data and primer sequences allows defendants to challenge the reliability of the kit results by, among other things: questioning whether the kits were validated appropriately, showing that the kits are prone to certain types of errors, and looking for the effects of contamination in the PCR process. Moreover, the defense would be able to provide the jury with more information on the quality of the DNA evidence provided to it, thus enhancing the fact finding process.

## C. Due Process

The constitutional clause most frequently discussed by courts in evaluating defense subpoenas for information held by private corporations is the Due Process Clause. In State v. Pfenning, for example, the trial court excluded the results of PowerPlex, Profiler Plus, and Cofiler where neither the defense nor the scientific community at large had access to validation data, stating "the manufacturer's proprietary concerns can not [sic] trump Defendant's right to a fair trial." Other trial courts have disagreed with the Pfenning court, holding that lack of access to the data does not raise any due process issues. In State v. Dishmon, the trial court found that PE's refusal to divulge primer sequences and developmental validation data did not violate the defendant's constitutional rights because the defendant "suffer[ed] no known harm by not having it." In so holding, the Dishmon court and others misunderstood the contours of the Due Process Clause.

In Brady v. Maryland, the Supreme Court held that the Due Process Clause of the Fourteenth Amendment requires the prosecutor to disclose evidence where the evidence is favorable to the defendant and "material either to guilt or to punishment, irrespective of the good faith or bad faith of the prosecution."

On first glance, Brady may appear to provide for a prosecutorial duty to disclose relevant evidence only where it is in the prosecutor's possession; that is, Brady seems not to apply to information held by third parties. However, in Kyles v. Whitley, the Supreme Court reversed a capital murder conviction for failure to disclose evidence even though that evidence was not in the possession of the prosecutor. The exculpatory evidence in question in Kyles was in the possession of the police. Instead of simply extending the Brady guarantee of disclosure to include information held by police officers, the Court went further:

> *The individual prosecutor has a duty to learn of any favorable evidence known to the others acting on the government's behalf in the case, including the police. But whether the prosecutor succeeds or fails in meeting this obligation (whether, that is, a failure to disclose is in good faith or bad faith), the prosecution's responsibility for failing to disclose known, favorable evidence rising to a material level of importance is inescapable.*

Under Kyles, the first question of a Brady analysis becomes: is the holder of the information requested by the defense "acting on the government's behalf in the case?" The question cannot be answered without a brief review of the history of development of DNA kits. When Congress authorized the creation of a National DNA Database in 1994, the FBI felt that the DNA testing methods then available were not sufficiently sophisticated to form the basis of a national standard. Dr. Budowle, the director of the Forensic Science Research and Training Center at the FBI Academy, has testified that he believed a multiplex PCR-based system would be the best system for national use. In 1994, the only multiplex systems available were triplex systems that he found to be inadequate. As a result, he decided to find a manufacturer that was willing to work with the FBI to develop a more complex multiplex system than those in existence at the time. He felt only two companies were capable of developing an appropriate DNA typing system: PE and Promega. Dr. Budowle chose to talk to PE first because the FBI and PE had worked together in the past, and PE kits already were used by the FBI and other state crime laboratories.

PE and the FBI together developed the new system over several years and through several phases. The first step required the development of the primer sequences to be used in PCR. The second and third steps involved developmental and forensic validation. During these stages, PE worked closely with the FBI and a group of laboratories chosen by Dr. Budowle. The laboratories tested early versions of the multiplex kit and reported the results of those tests to PE. In 1998, the results of the collaboration, the Profiler Plus and Cofiler kits, went on the market.

Any claim that PE developed its multiplex kits in an entirely private sphere, independent of government involvement, is disproved by the history of its development. The government, in the form of FBI scientist Dr. Budowle, both suggested the form that the kits should take and participated extensively in their development. Nonetheless, this history does not entirely support a finding that the kit makers work "on behalf of the government." Undoubtedly PE acted on its own behalf as well. Moreover, other kit makers, such as Promega, may have no such ties to the government.

As a result, it is imperative to consider the role of the kit makers in the larger context of their role in the criminal justice system. Under state action law, it is clear that "when private individuals or groups are endowed by the State with powers or functions governmental in nature, they become agencies or instrumentalities of the State." Under this rule, private individuals or corporations become agencies or instrumentalities of the state if they run elections, manage a company town or maintain a municipal park. In the 1970s, the Court narrowed the class of private entities considered agencies of the government to those that exercise powers traditionally and exclusively reserved to the government.

Even under this narrowed test, the performance of scientific testing in the course of police investigation is an activity traditionally and exclusively reserved to the government. No area is more firmly within the exclusive and traditional sphere of activity of the state than law enforcement. West v. Atkins, a case involving prisons, is particularly instructive. In West, the Court held that a private physician under contract with North Carolina to provide medical services at a state prison hospital on a part-time basis was a state actor. The Court noted that because provision of medical care to prisoners was fundamentally a governmental activity, the provider of that care is a state actor. Moreover, this analysis is not altered by the fact that respondent was paid by contract and was not on the state payroll, nor by the fact that respondent was not required to work exclusively for the prison.

Like the prison hospital, the crime laboratory analyzing evidence in the course of a criminal investigation is performing a law enforcement function exclusively associated with the government. Corporations that work hand in hand with those laboratories by providing the technology and know-how to perform DNA testing are, like the doctor in West, performing a fundamentally governmental "function within the state system." As a result, those corporations must be considered to be "acting on the government's behalf in the case" within the meaning of Brady.

Aside from the historical and state action arguments, it is worth noting that several lower courts implicitly have found that private corporations act on behalf of the government when they perform DNA testing on behalf of law enforcement. For example, in People v. Davis, a conviction was reversed because Lifecodes, a private laboratory, refused to disclose statistical standards used in determining the probability of a "match." The court described the evidence as "`material either to guilt or punishment'" and cited to Brady. The court then held:

> *"The rule is plain that where the prosecution is permitted to call a witness, expert or not, who testifies as to a fact in issue or a conclusion to be drawn, the defendant is entitled to examine the underlying data, the basis for the testimony" or else the defendant suffers undue prejudice.*

The Minnesota Supreme Court also has held that due process rights are implicated where information relating to testing is not revealed by private laboratories. Although these cases differ from the main concern of this Note – information underlying kits made by private corporations but used by state and other laboratories –

they demonstrate the degree to which forensic DNA testing is considered government activity. Given the historical and state action arguments suggesting that kit makers are acting on behalf of the government, what is the prosecution's duty with respect to Brady evidence held by them? In Kyles v. Whitley, the Court framed the prosecutor's duty in the following terms:

> *No one doubts that police investigators sometimes fail to inform a prosecutor of all they know. But neither is there any serious doubt that "procedures and regulations can be established to carry [the prosecutor's] burden and to insure communication of all relevant information on each case to every lawyer who deals with it."*

The Supreme Court thus insisted that the prosecutor "establish regulations and procedures" for obtaining information from other law enforcement agencies, which, as was established above, should be understood to include private forensic technology companies that participate in testing. But the Supreme Court seemed to say more than that the prosecutor has a duty to search for the relevant information; the Supreme Court assumed that these regulations and procedures would enable the prosecutor to discover the information. As a result, the Supreme Court created no room for failure. The core inquiry made by a Court hearing a Brady claim does not relate to the prosecutor's conduct; it relates to the "character of the evidence that [the prosecutor] has withheld."

What "character" defines Brady evidence? Evidence must be disclosed under Brady where it is "material either to guilt or punishment." The Supreme Court has identified two concerns underlying materiality of evidence: whether the evidence could have changed the outcome of the trial, and whether the court could make an accurate and fair ruling without the evidence.

The Court described the first concern in United States v. Bagley, where it wrote that "evidence is material only if there is a reasonable probability that, had the evidence been disclosed to the defense, the result of the proceeding would have been different. A `reasonable probability' is a probability sufficient to undermine confidence in the outcome." Where information sought by the defense could result in evidence against the defendant being ruled inadmissible, that information usually will have "a reasonable probability" of altering the outcome of the case. In United States v. Yee, the district court addressed the materiality of data that challenges the admissibility of DNA evidence:

> *If [defendants] prevail in their challenge to the government's proffer of that evidence, the likelihood of conviction will be diminished substantially, if not eliminated. The opportunity for the defendants to make an effective challenge, and, more importantly, the ability of this court to resolve the issue of admissibility fairly, depend on prehearing access by defense counsel and their experts to the materials that are the subject of their discovery request.*

Although the probability that the exclusion of DNA evidence would have changed the course of the trial will vary from case to case, there are certainly many cases in which the prosecution relies heavily on DNA evidence. In most cases in which the evidence is present, it forms a strong part of the prosecution's case-in-chief. As a result, information that bears upon the admissibility of that evidence must be considered capable of altering the outcome of the trial. The Yee court also referred to the second concern underlying the materiality requirement: the ability of courts to find the truth. The Court's most recent word on the subject came in Kyles v. Whitley, where the Court emphasized, "The question is not whether the defendant would more likely than not have received a different verdict with the evidence, but whether in its absence he received a fair trial, understood as a trial resulting in a verdict worthy of confidence." The possibility of a fair trial is certainly reduced where information related to the value of scientific evidence is withheld from the defense. In its order granting discovery to the defense, the Yee court wrote, "The precondition to reaching a correct decision concerning the admissibility of the DNA evidence in this case is a record that is as complete and clear as is reasonably possible." Where creating a complete and clear record requires the prosecution to provide evidence in the possession of law enforcement, under Brady the prosecution must do so.

Finally, requiring the prosecutor to obtain this information from the sellers of the kits serves the fundamental goal of due process: ensuring a fair trial. Requiring disclosure is the only way to ensure that private corporations do not play too great a role in the administration of justice. In Kyles the Court wrote: "any argument for excusing a prosecutor from disclosing what he does not happen to know about boils down to a plea to substitute the police for the prosecutor, and even for the courts themselves, as the final arbiters of the government's obligation to ensure fair trials." The role of the courts as the arbiters of the government's obligation to ensure fair trials is equally imperiled by private corporations.

## Conclusion

As scientific evidence becomes more prominent in the courtroom, subpoenas to corporations in possession of scientific information are likely to issue with greater frequency. This conflict promises especially grave consequences in the area of criminal prosecutions, where science may be used to secure a sentence of imprisonment or death. Although courts desire to preserve a space for innovation in forensic and other sciences likely to increase the efficacy of the criminal justice system, this need must be balanced against the rights of defendants to challenge scientific evidence against them. In the end, the intellectual property rights of corporations must give way to the constitutional rights of the accused.

# 9

# Genetic Discrimination and the Workplace

## Introduction

The ability of employers to literally look beyond the skin of their employees to learn what were heretofore DNA secrets was recently signaled at a gala White House ceremony. Joined by Francis Collins, the Director of the International Human Genome Project, and J. Craig Venter, President of Celera Genomics, President Clinton celebrated the "completion of the first survey of the entire human genome." An international consortium of scientists had completed a "map of overlapping fragments that includes 97% percent of the human genome." Deemed a "stunning, humbling achievement," this research holds the promise of revolutionizing "the diagnosis, prevention and treatment of most, if not all, human diseases." It has the potential to be used to: predict whether a person is likely to suffer from a particular disease; tailor drug prescriptions for optimum efficiency for each individual; alter a patient's genetic material to prevent or fight disease; prevent genetic defects from being transferred to future generations; and enhance understanding of human evolution. Inevitably, the sweeping scope of this information also raises questions about how access to genetic information will affect the workplace. The ultimate question concerns the balance between an employee's right to privacy and an employer's need to know what may be the utmost in sensitive private genetic information.

As with any advancement in science and technology, questions about how such information can be used in less than life-enhancing ways immediately arise. President Clinton was quick to underscore the importance of the project, while also acknowledging the accompanying concerns:

> *We must not shrink from exploring that far frontier of science. But as we consider how to use new discovery, we must also not retreat from our oldest and most cherished human values. We must ensure that new genome science and its benefits will be directed toward making life better for all citizens of the world, never just a privileged few. As we unlock the secrets of the human genome, we must work simultaneously to ensure that new discoveries never pry open the doors of privacy. And we must guarantee that genetic information cannot be used to stigmatize or discriminate against any individual or group.*

The two scientists most responsible for the genome sequence and joining the President at the gala reiterated his sentiments. Defying the words of DNA scientist James Watson that, "in large part, our fate is our genes," Dr. Collins took a clearly anti-essentialist position. He explained that we are "clearly much, much more than the sum total of our genes. our physiology is based on complex and seemingly infinite interactions amongst all our genes and the environment, just as our civilization is based on the interactions amongst all of us." Collins also called for "new laws to protect us from genetic discrimination to maximize the medical benefits from genome discoveries." Similarly, Dr. Venter stated: "I am concerned, as many of you are, that some will want to use this new knowledge as a basis of discrimination.... While those who will base social decisions on genetic reductionism will ultimately be defeated by science, new laws to protect us from genetic discrimination are critical in order to maximize the medical benefits from genomic discoveries." Those closest to the genome project have, at the least, endeavored to raise public awareness about the clear inadequacy of current laws, both federal and state, to address issues raised by genetic research and to prevent discrimination on the basis of predictive medical information. The words of Sir Isaac Newton about movement in the physical world "To every action there is always opposed an equal reaction" portend the pernicious downside to genetic discoveries. The same information that can be used to "alert patients that they are at risk for certain diseases" and to precisely diagnose and reliably predict the course of disease, could also be used to deny individuals equal opportunity in employment and insurance coverage.

The review of current law demonstrates that comprehensive federal legislation is needed in order to prohibit discrimination on the basis of predictive genetic information and to provide uniform protection for individuals across the United States. The most effective legislation for the workplace would protect the privacy of genetic information; prohibit discrimination by employers in hiring, promotion, firing, and other employment decisions; prohibit discrimination by insurers; yet also provide employers with information which is necessary to maintain a safe workplace. The proposed Genetic Nondiscrimination in Health Insurance and Employment Act of 1999 appears to address these issues.

## Mapping the Human Genome: Overview

The Human Genome Project is an international research program established "to construct detailed genetic and physical maps of the human genome, to determine the complete nucleotide sequence of human DNA, to localize the estimated 50,000-100,000 genes within the human genome, and to perform similar analyses on the genomes of several other organisms." Often referred to as the "Book of Life," such a map could serve as a "source book for human biology and medicine for centuries to come," which will "facilitate greatly the elucidation of genetic disease." The information that can be revealed by genetic tests includes "presymptomatic medical information about an individual, including information about an individual's increased risk of future disease, disability, or early death." Moreover, such tests can also reveal "information about an individual's carrier status, that is, the likelihood of parents passing on to their children a genetic condition, and about the health of the individual's family members."

Dr. Venter actually envisions a future in which anyone can get a "DVD" complete with his or her genetic code. At this early stage, scientists have demonstrated that "errors" inherited in our genes are responsible for an about 3,000 to 4,000 diseases, including Huntington's disease, neurofibromatosis, cystic fibrosis, and Duchenne muscular dystrophy. New and emerging genetic technology is designed to determine if people have an increased likelihood of developing a particular disorder. One such example is the identification of a genetic alteration that appears to double a person's risk of colon cancer. A $200 blood test can be used to identify the mutation; people so identified can then use regular colon examinations to detect early cancer growth when it is most readily treated. However, because "interaction of genes with environmental factors, including occupation, diet and lifestyle," affect whether a disease will actually manifest itself, discretion should be used in evaluating genetic data.

In any endeavor of this scope and significance, concerns are raised about the moral implications of such discoveries. Accordingly, the U.S. Department of Energy and the National Institute of Health have "devoted 3% to 5% of their annual Human Genome budgets toward studying the ethical, legal, and social issues (ELSI) surrounding the availability of genetic information." Nearly a dozen broad categories of concern have been identified:

- Fairness in the use of genetic information by insurers, employers, courts, schools, adoption agencies, law enforcement, and the military.
- Privacy and confidentiality of genetic information.
- Psychological impact and stigmatization due to an individual's genetic differences.
- Genetic testing of an individual for a specific condition due to family history (prenatal, carrier, and presymptomatic testing) and population screening (newborn, premarital, and occupational).
- Reproductive issues, including informed consent for procedures, use of genetic information in decision making, and reproductive rights.
- Gene therapy which may someday be used to treat, cure, or prevent genetic disorders.
- Genetic enhancement, including gene therapy to supply a characteristic such as height, that a parent might want in a

child but which does not involve the treatment or prevention of a disease.

- Fairness in the use of genetic technologies (including access and payment for these expensive technologies).
- Clinical issues, including education health service providers, patients, and the general public, and implementation of standards and quality-control measures in testing procedures.
- Commercialization of products raising property rights issues (patents, copyrights, and trade secrets) and accessibility of data and materials.
- Conceptual and philosophical implications regarding human responsibility, free will versus genetic determinism, and concepts of health and disease.

As DNA banks begin to proliferate, these issues need to be addressed to ensure that the creation, maintenance, and use of DNA data obtained from individuals is sufficiently protected. A perfect example of this need can be found in the case of Mayfield v. Dalton. There, the constitutionality of a Department of Defense program, which collects and stores blood and tissue samples from members of the armed forces for future DNA analysis, was challenged.

Marine Corps soldiers John C. Mayfield, III and Joseph Vlacovsky argued that the military's "compulsory taking of specimens without proper safeguards to maintain the privacy of the donor was a violation of the Fourth Amendment prohibition against unreasonable searches and seizures." Attempting to bring a class action on behalf of all military personnel who provided blood and tissue samples, plaintiffs also argued that they "feared that information obtained from the repository samples regarding the donors' propensities for hereditary diseases and genetic disorders might be used to discriminate against applicants for jobs, insurance, or benefit programs." The government defended the program, asserting that the purpose of collecting and storing the samples in a DNA repository was to identify soldiers' remains.

Inasmuch as Mayfield and Vlacovsky were honorably separated from military service while the case was pending on appeal and were no longer subject to the DNA collection program, the Ninth Circuit rejected their challenge, vacated the judgment of the district court, and remanded the case with instructions that it should be dismissed as moot. The litigation prompted the military to change the repository in ways that appear to address the concerns raised. For example, as of April 1996, the maximum length of time that the specimens may be retained was shortened from seventy years to fifty. Additionally, the military will now destroy individual specimens, upon the request of the donor, following the conclusion of the donor's military obligation." The issue regarding the use of stored genetic data, however, remains unresolved. Similarly, the other ELSI identified are open to "real world" controversy. In particular, one critical arena that needs to be addressed is the workplace.

## Workplace Implications: Genetic Monitoring and Genetic Screening

Genetic testing in the workplace is characterized by genetic monitoring and genetic screening. Genetic monitoring "involves periodically examining employees to evaluate modifications of their genetic material—e.g., chromosomal damage or evidence of increased occurrence of molecular mutations—that may have evolved in the course of employment." Such testing is often conducted to monitor workplace exposure to hazardous substances, which could indicate increased likelihood of future illness. In general, genetic monitoring is performed on groups of employees. It seeks to identify risks for an exposed group as a whole, to target work areas for increased health and safety precautions, and to indicate a need to reduce exposure levels for a group exposed to previously unknown hazards.

In contrast, genetic screening involves tests "to examine the genetic makeup of employees or job applicants for certain inherited characteristics." One use of genetic screening is to determine if employees or job applicants have "genetically determined traits that render them susceptible to a pathological effect if exposed to specific agents." In other words, genetic screening assesses whether or not a person has a "genetic predisposition to an occupationally related disease." The other use of genetic screening is not necessarily employment-related. Employees and job applicants might be tested to detect "general heritable conditions," that is, genetic predisposition to disease that might make them less productive and costly for health insurance purposes.

Many workers fear that genetic information gathered in monitoring and screening will be used unfairly by employers to form the basis of hiring, promotion, termination, and health care decisions. Relatively recent history illustrates that there may be cause for worry. During the 1970s, many leaders in the Black community felt that sickle cell anemia was not receiving adequate medical attention. As a result of increased interest in this disease, Congress passed the National Sickle Cell Anemia Control Act of 1972 to appropriate federal funds for voluntary testing programs. Sickle cell screening laws were also passed in at least twenty states, targeting "newborns, schoolchildren, marriage license applicants, and inmates at penal institutions." Individuals were screened for sickle cell anemia (when the patient inherits the gene for sickle hemoglobin from both parents) and sickle cell trait (when the gene for sickle hemoglobin is passed from only one parent). Because of a lack of adequate genetic counseling, however, a great deal of confusion was generated about how to interpret the test results. This was further complicated by the fact that some scientists suggested that "carders of the gene might be hypersusceptible to certain workplace toxins such as benzine, lead, cadmium, carbon monoxide, and cyanide." Reverberations in the workplace were soon to follow. Some individuals who were found to be "carriers of the sickle cell trait experienced discrimination at work and from insurance companies that raised their premiums." Additionally, some job applicants were denied employment, while others were terminated from their jobs because they were found to be sickle cell carriers. This stigmatization and discrimination was further exacerbated by a lack of confidentiality. Thus, what began as a public health initiative degenerated into a means of stereotyping and discriminating against African-Americans.

Screening for sickle cell anemia was followed by reports of increasing genetic monitoring and screening. This prompted the House Committee on Science and Technology to hold hearings during the late 1970s and early 1980s and to request that the Office of Technology Assessment conduct an investigation. The National Opinion Research Center (NORC) conducted a survey for the OTA from February 25 to June 8, 1982, by sending "confidential questionnaires to the chief executive officers of the 500 largest U.S. industrial companies, the chief executive officers of the 50 largest private utility companies, and the presidents of the 11 major unions that represent the largest numbers of employees in those companies." The result: "Of the 366 organizations responding, 6 companies (1.6 percent) were currently using genetic testing, 17 (4.6 percent) used some tests in the past 12 years, 4 (1.1 percent) anticipated using the tests in the next 5 years, and 55 (15 percent) stated that they would possibly use the tests in the next 5 years." The overall conclusion tended to downplay the significance of the results, stating that genetic testing was being used by a "few companies, that its use has declined in the past 12 years, but that it may be used by more companies in the future." These results were, however, significant enough to prompt concern about workplace implications.

In light of the quick progress in both occupational medicine and human molecular genetics, the OTA reexamined the use of genetic testing in the workplace. From March 24 to July 15, 1989, a follow-up survey was conducted by the OTA Consistent with the 1982 survey, the 1989 survey addressed the 500 largest U.S. industries (Fortune 500), the fifty largest utilities, and thirty-three major labor unions. The updated survey was designed to "remove ambiguities that might have been present in the initial survey" and "to include a representative sample of all other companies with 1,000 or more employees so that broader estimates could be made of the use and pattern of genetic monitoring and screening in the workplace." In the 1989 survey, "12 health officers (5 percent) reported their companies currently conducted genetic monitoring or screening (1 of the companies used genetic monitoring and screening while 11 used only genetic screening)." Based on this data, the OTA concluded that there has been "little or no growth in the number of companies conducting genetic monitoring and screening in the workplace." Like the 1982 survey, the 1989 survey also elicited responses about anticipated future use of genetic medical information. The conclusion: although the literal numbers "cannot be directly compared to the current survey, the 1989 OTA survey appears to indicate fewer companies anticipate using genetic monitoring or screening."

Despite these findings, which suggest that medical testing may not be widespread, the 2000 American Management Association survey on medical testing in the workplace found that "seventy-one percent of major U.S. firms require medical examinations of new hires, current employees, or both" and "40 percent of the respondent firms require unscheduled exams when job performance suggests a medical problem." This data is based on "2,133 usable responses" to questionnaires that were mailed in January 2000 to human resource managers in AMA member and client companies. With regard to genetic testing, the study was designed to elicit more explicit and detailed responses than previous surveys. When presented with the following definition of "genetic testing," seven respondents answered that their firms performed genetic testing:

> *The analysis of human DNA, RNA, chromosomes, proteins, and certainmetabolites in order to detect inheritable disease related to genotypes, mutations, phenotypes, or karyotypes for clinical purposes. Such purposes include predicting risk of disease, identifying carriers, and establishing prenatal and clinical diagnosis or prognosis.*

A much larger percentage— 15.8% – reported testing for "Susceptibility to Workplace Hazards," suggesting that, even though all such testing is not genetic, the definition may have skewed the responses. Perhaps even more importantly, forty-eight percent of employers revealed that they conduct medical testing under the broad category "Fitness for Duty," defined as "establishing the ability of the applicant or employee to perform assigned job tasks." Theoretically, this could include genetic testing to determine whether a person is at risk for a disease which could affect his or her fitness. Moreover, 49.2% of the time, the results of this testing were used for some purpose in the employment context. Because of the way the data are grouped, however, it is not possible to ascertain if the action taken was adverse. The AMA findings regarding susceptibility and fitness, as well as other specific genetically-related categories are as follows:

| TESTING CATEGORIES | Any such Testing Hire | New Hire Testing | Employee Testing | Results Used to job applicants |
|---|---|---|---|---|
| Fitness for Duty | 48.0% | 40.9% | 25.8% | 42.1% |
| Susceptibility to Workplace Hazards | 15.8% | 12.4% | 11.4% | 8.5% |
| Breast/Colon Cancer | 3.1% | 1.3% | 2.7% | 0.7% |

(Contd.)

| TESTING CATEGORIES | Any such Testing Hire | New Hire Testing | Employee Testing | Results Used to job applicants |
|---|---|---|---|---|
| HIV Infection | 2.6% | 2.1% | 0.7% | 1.5% |
| Sickle Cell Anemia | 0.9% | 0.8% | 0.5% | 0.9% |
| Pregnancy | 0.8% | 0.6% | 0.4% | 0.7% |
| Huntington's Disease | 0.4% | 0.3% | 0.2% | 0.5% |
| Family Medical Histories | 18.1% | 15.3% | 10.1% | 4.0% |

| TESTING CATEGORIES | Results Used to Assign or reassign employees | Results Used to Retain or dismiss employees | Results used in any regard |
|---|---|---|---|
| Fitness for Duty | 23.5% | 14.3% | 49.2% |
| Susceptibility to Workplace Hazards | 8.6% | 2.8% | 12.2% |
| Breast/Colon Cancer | 0.8% | 0.1% | 1.6% |
| HIV Infection | 1.0% | 0.1% | 2.3% |
| Sickle Cell Anemia | 0.7% | 0.1% | 1.5% |
| Pregnancy | 1.7% | 1.7% | 3.8% |
| Huntington's Disease | 0.3% | 0.1% | 1.1% |
| Family Medical Histories | 1.5% | 0.4% | 4.9% |

This study shows that at least some employers are in fact currently using the results of genetic testing and other health-related medical testing in connection with employment decisions. Accordingly, the AMA reiterated its advisory to member companies regarding the use of medical information:

> *Genetic testing allows employers access to intimate personal information about employees and therefore demands the highest standard of responsibility in its exercise. The results of medical tests of any type should be confidential information available only to the*

*employee and the appropriate human resources and medical officers of the organization.*

Employees and job applicants who are subject to genetic testing should be fully informed of its purpose. They should know in specific detail what the test is intended to find out and the effect of a positive test on their employment status. They should be informed of their test results in clear language. Employees who are physically able to perform job tasks, and whose health is not endangered by performing them, should be free to apply for such positions and to retain them as long as they meet performance standards.

Even though the AMA recommends that employers inform their employees about the results of any medical tests, firms responding to the 2000 survey admitted that 16% of job applicants and 11.5% of current employees are not notified.

Inasmuch as the AMA study reveals that many employers are engaging in medical testing and that much of this testing is conducted using blood samples and scrapings from inside the cheek, employers have the potential to surreptitiously test for a wide and ever-growing variety of medical conditions. This is precisely what prompted the litigation in Norman-Bloodsaw v. Lawrence Berkeley Laboratory. The plaintiffs, present and former employees of Lawrence (a research institution jointly operated by state and federal agencies), alleged that "in the course of their mandatory employment entrance examinations and on subsequent occasions, Lawrence, without their knowledge or consent, tested their blood and urine for intimate medical conditions - namely, syphilis, sickle cell trait, and pregnancy." The complaint asserted that this testing "violated Title VII of the Civil Rights Act of 1964, the Americans with Disabilities Act , and their right to privacy as guaranteed by both the United States and California Constitutions." The United States District Court for the Northern District of California granted the defendant's motions for dismissal, judgment on the pleadings, and summary judgment on all of the plaintiffs' claims. The Ninth Circuit, however, affirmed as to the ADA claims, but reversed as to the Title VII and state and federal privacy claims, stating:

*Because material and disputed issues of fact exist with respect to whether reasonable persons in plaintiffs' position would have had reason to know that the tests were being performed, and because the*

> *tests were a separate and more invasive intrusion into their privacy than the aspects of the examination to which they did consent, the district court erred in granting summary judgment on statute of limitations grounds with respect to the Title VII claims and the federal and state privacy claims.*

Moreover, the Ninth Circuit held that the district court also "erred in dismissing the federal and state privacy claims and the Title VII claims on the merits." In reaching its decision in Norman-Bloodsaw, the Ninth Circuit stated: "One can think of few subject areas more personal and more likely to implicate privacy interests than that of one's health or genetic makeup. The conditions tested for were aspects of one's health in which one enjoys the highest expectations of privacy."

> *The court also found that to a general medical examination does not abolish one's privacy right not to be tested for intimate, personal matters concerning one's health—nor does consenting to giving blood or urine samples, or filling out a questionnaire. As we have made clear, revealing one's personal know-ledge as to whether one has a particular medical condition has nothing to do with one's expectations about actually being tested for that condition.*

Sustaining plaintiffs' Title VII claims, the court determined that the claims fell neatly into the Title VII framework. The plaintiffs had alleged that black and female employees were singled out for added nonconsensual testing, and therefore, defendants had selectively invaded the privacy of certain employees on the basis of race, sex, and pregnancy. More recently, the EEOC filed a petition for a preliminary injunction against Burlington Northern Santa Fe Railroad, seeking to end genetic testing of all employees who have filed claims for work-related injuries based on carpal tunnel syndrome. This was the first suit by the EEOC challenging genetic testing. The EEOC alleged that the employees were not told of the genetic test or asked to consent to it. Moreover, in at least one instance, an employee who refused to provide a blood sample because he suspected that it would be used for genetic testing was allegedly threatened with imminent discharge if he failed to submit the sample. Burlington Northern asked employees to submit to blood tests to see if they carried a genetic trait called Chromosome 17 deletion; some studies suggest that a person with that trait is more likely to develop some forms of carpal tunnel syndrome. Within days of their request,

Burlington Northern agreed to stop requiting genetic testing of employees who file claims for carpal tunnel syndrome.

The court challenges to genetic monitoring and screening are just beginning, but evidence of genetic discrimination is apparent in a 1996 study conducted by a joint effort, which included researchers at Harvard Medical School and Stanford School of Medicine. The results showed that a number of institutions, including health and life insurance corporations, clinical professionals, adoption agencies, armed services, employers, educational institutions, and blood banks, have engaged in such discrimination. Specifically with regard to employers, the study revealed that in

> *many of the cases involving employment, individuals believed that they were not hired or were fired because they were at-risk for genetic conditions. In other cases, individuals who were employed reported that they were reluctant to seek either a more desirable job or a job in a different location because they feared that they would be unable to obtain health insurance in their new position.*

Two case illustrations are used to depict genetic discrimination. In the first scenario, a twenty-four-year-old woman was "fired from her job as a social worker shortly after her employers learned that she was at-risk to develop Huntington disease. In the eight-month period prior to her termination, she received three promotions and outstanding performance reviews." After revealing that a family member has Huntington's disease, she "was given a poor performance review," with no examples of poor performance. Thereafter, she was fired and told by a co-worker that the "employer was concerned about her risk to develop Huntington disease." In another employment-related case, during a job interview, a fifty-three-year-old man disclosed that he was diagnosed with asymptomatic hemochromatosis. In his second interview, he was told that the company was interested in hiring him, but would not be able to offer him medical insurance; during a third interview he was told that "although they would like to hire him, they were unable to do so because of his hemochromatosis." The overall conclusion is that "although not systematic, genetic discrimination does occur in a wide variety of contexts and can cause hardship to affected individuals and their families," and that because "the number and use of genetic tests is expanding rapidly and will continue to increase, it is vital that standards be developed in the near future to ensure that genetic information be used fairly." Informal anecdotal evidence of workers being denied jobs, fired, and losing insurance coverage further reinforces these conclusions.

Yet, notwithstanding the potential for and evidence of misuse, "employers may have legitimate business reasons for obtaining employees' medical information," including the results of genetic testing. For example, employers need to comply with safety laws, such as OSHA. For this reason, employees' right to privacy and employers' need to know must be balanced to achieve "a system that allows the nondiscriminatory use of genetic information by employers while affording the employee/applicant protection against the misuse and dissemination of genetic information."

## Existing Protection against the Discriminatory Use of Genetic Information in the Workplace

At the federal level, there is some protection against discriminatory use of genetic information in the workplace. The existing protection, however, falls short of the comprehensive law needed to protect the rights of workers and to clarify the appropriate use of genetic information for employers.

### A. Executive Order

On February 8, 2000, President Clinton signed Executive Order 13145 to prohibit discrimination in federal employment based on genetic discrimination. Speaking before the American Association for the Advancement of Science (AAAS), President Clinton stated:

> *Fear of misuse of private genetic information is already very widespread in this nation. Americans are genuinely worried that their genetic information will not be kept secret, that this information will be used against them. As a result, they are reluctant to take advantage of new breakthroughs in genetic testing. If we do not protect the right to privacy, we may actually impede the reach of these breakthroughs in the lives of ordinary people, which would be a profound tragedy.*

The purpose of the Executive Order is "to provide equal employment opportunity in Federal employment for all qualified persons and to prohibit discrimination against employees based on protected genetic information, or information about a request for the receipt of genetic services." The Order, which applies to all federal current and former civilian workers, as well as applicants for federal jobs, defines "protected genetic information" as: information about an individual's genetic tests; information about the genetic tests of an individual's family members; or information about the occurrence of a disease, or the medical condition or disorder in family members of the individual.

Information about "an individual's current health status (including information about sex, age, physical exams, and chemical, blood, or urine analyses) is not protected genetic information" unless it falls within one of the above categories. Another important aspect of the Order is that the employing agency or department is directed to treat such information as "confidential medical records" separate from general personnel files.

Even though the Executive Order does not create any legally-enforceable right or benefit, because "much of what is prohibited under the Executive Order is also prohibited under section 502 of the Rehabilitation Act of 1973," the Equal Employment Opportunity Commission (EEOC) advises that "federal sector applicants and employees who believe that a department or agency has violated a provision of the Executive Order may pursue that issue under the familiar procedure set forth at 29 C.F.R. Part 1614." Under this procedure, the first step in the process is to "contact an EEO [Equal Employment Opportunity] counselor within 45 days of the alleged violation." Thereafter, the process involves multiple steps of administrative review and investigation.

In signing the Order, President Clinton said that his "goal is to set an example and pose a challenge for every employer in America, because no employer should ever review your genetic records along with your resume." AAAS Executive Officer Dr. Richard S. Nicholson enthusiastically agreed, asserting that "genetic information should be used to enhance, not undermine, the quality of life for individuals, and for society" and urging the "adoption of policies to prevent misuse of genetic testing results."

## B. Americans with Disabilities Act

The Americans with Disabilities Act of 1990 may prohibit discrimination based on genetic information, yet there is no definitive judicial ruling on this point. Enacted just over ten years ago, the employment provisions for employers with twenty-five or more employees became effective on July 26, 1992; beginning on July 26, 1994, the ADA became applicable to employers with fifteen or more employees. The right to money damages available to private sector employees under the ADA, however, does not extend to state employees.

The ADA prohibits discrimination against "a qualified individual with a disability," meaning "an individual with a disability who, with, or without reasonable accommodation, can perform the essential functions of the employment position that such individual holds or desires." To establish prima facie case of discrimination under the ADA, the plaintiff must show by a preponderance of the evidence that the employer is subject to the ADA, the plaintiff is "disabled" within the meaning of the ADA, that plaintiff is otherwise qualified to perform the essential functions of the job, with or without reasonable accommodation, and plaintiff suffered adverse employment action because of the disability. In the context of genetic discrimination, one must determine if a genetic condition (manifested or asymptomatic) is considered to be a "disability" under the ADA. Under the ADA, disability is defined as:

> *A physical or mental impairment that substantially limits one or more of the major life activities of such individual;*
>
> *A record of such impairment; or Being regarded as having such an impairment.*

Like any other disability, the ADA is designed to protect individuals who "have a genetically related illness or disability once it becomes manifest and substantially limits a major life activity." Note, however, that "if a person is taking measures to correct for, or mitigate, a physical or mental impairment, the effects of those measures – both positive and negative -- must be taken into account when judging whether that person is substantially limited' in a major life activity and thus `disabled' under the Act." Accordingly, a person could be deemed to have a disability under the first definition if, notwithstanding the use of a corrective device, that individual is substantially limited in a major life activity. This is to say that an individualized inquiry must be made to determine whether a person has a disability based on the person's actual condition.

Under the rationale set forth in Sutton v. United Air Lines, courts must look to the person's actual condition to determine if a person with a genetic condition falls under the first definition of a disability under the ADA. For example, a person with a genetic propensity for Parkinson's disease would not be "disabled" under the Act merely by virtue of general information about how this condition usually affects individuals. Such a person should be treated as an individual, not as a member of a group with similar impairments. Thus, a person with Parkinson's disease whose illness does not impair his or her daily activities would not necessarily be considered disabled.

A genetically-related illness that has already manifested itself can be analyzed under established case law to determine if it substantially limits one or more major life activities. What is much more uncertain, however, is whether the ADA restricts discrimination when a genetic condition or trait is asymptomatic. Such a situation could arise, for example, when an employer makes a conditional offer of employment that requires the applicant to submit to a medical exam. If genetic testing reveals that this otherwise healthy woman has the gene for Huntington's Disease, the potential employer might withdraw the offer based on fear about the course this degenerative neurological disease could take, which could affect her productivity and cause the company to incur substantial insurance costs. Under the ADA, once a conditional offer is extended to a job applicant, there are no restrictions on the extent of an employer's medical examination; an employer may require applicants to disclose their medical history, including medical tests. In the case of current employees, questions exist about whether an employer can use genetic testing in connection with some job-related reason, such as job placement. What if an employer learns about a genetic condition which it perceives as an impairment?

Under such circumstances, the question is whether the genetic condition or trait falls within the definition of a disability because it is regarded as an impairment by an employer. Individuals may fall within this statutory definition if an employer (who is covered by the ADA) mistakenly believes that a person has a physical impairment that substantially limits one or more major life activities, or mistakenly believes that an actual, non limiting impairment substantially limits one or more major life activities. As the Supreme Court stated in Sutton v. United Air Lines, "it is necessary that a covered entity entertain misperceptions about the individual it must believe that one has a substantially limiting impairment that one does not have or that one has a substantially limiting impairment when, in fact, the impairment is not so limiting." Such misperceptions can be the result of stereotypical assumptions that are not based on an individual's ability. Accordingly, the attitudes of others toward an employee may give rise to a perceived disability.

Anticipating such situations, the Equal Employment Opportunity Commission (EEOC) issued interpretive guidelines about how to construe the language "regarded as having a substantially limiting impairment." In general terms, the EEOC underscores the fact that the ADA "also covers persons whose impairments are substantially limiting only as the result of the attitudes of others toward the impairment and persons who have no impairments but nonetheless are treated as having substantially limiting impairments." As the guidelines explain, the inclusion of persons regarded as having a substantially limiting impairment reflects Congressional intent to protect all persons who are subjected to discrimination based on disability, even if they do not in fact have a disability. It also reflects a recognition by Congress that the reactions of others to an impairment or perceived impairment can be just as disabling as the limitations caused by an actual impairment. Quoting a House Judiciary report, the EEOC explains that the definition of the term "disability" is "designed to protect against myths, fears, stereotypes, and other attitudinal barriers about disability" including, but not limited to, "'concerns about productivity, safety, insurance, liability, attendance, cost of accommodation and accessibility, and acceptance by co-workers and customers."' In short, these are precisely the kinds of concerns that individuals with genetic defects fear will be expressed against them in the workplace.

As the ADA legislative history specifically states, the "perception of the covered entity is a key element of this test." This rationale is probably the basis for the EEOC's interpretive guideline which states that the "regarded as having a substantially limiting impairment" aspect of the definition of disability:

> *applies to individuals who are subjected to discrimination on the basis of genetic information relating to illness, disease, or other disorders. Covered entities that discriminate against individuals on the basis of such genetic information are regarding individuals as having impairments that substantially limit a major life activity. Those individuals, therefore, are covered by the third part of the definition of "disability."*

To illustrate the point, the interpretive guidelines use an example in which a job applicant's genetic profile "reveals an increased susceptibility to colon cancer." The individual is "asymptomatic and may never in fact develop colon cancer," yet when the employer learns about the increased susceptibility to colon cancer, the conditional offer of employment is withdrawn, "because of concerns about matters such as [the potential employee's] productivity, insurance costs, and attendance." The EEOC 's position is that the employer is treating the applicant "as having an impairment that substantially limits a major life activity;" accordingly, the job applicant would be "covered by the third part of the definition of `disability.'" It should be remembered, however, that "people must show not only that they have a genetic defect, but also that they were regarded as disabled and discriminated against by an employer on account of that perception." The following may be evidence that an employee is regarded as having an impairment:

> *1) statements by the employer; 2) the employer's awareness of past medical history; 3) evidence of prejudice or concern about third parties' prejudices; 4) the employer's opinion that the plaintiff suffers from a specific condition, which, if true, would disqualify the plaintiff from the position; 5) concern about the employee's worker's compensation history or concern about the effect an employee may have on insurance premiums; 6) offering or providing accommodations, reassignment, disability benefits, light duty or medical leave to an employee; 7) stereotypes about a non disabiling condition; 8) the employer's failure to distinguish between a disability and a lack of qualifications; and 9) reliance on medical reports reflecting a serious impairment.*

Assuming that a prima facie case for genetic discrimination can be brought pursuant to the ADA, employers would have the full range of defenses currently available under the ADA to defend against such a charge, including "that an alleged application of qualification standards, tests, or selection criteria that screen out or tend to screen out or otherwise deny a job or benefit to an individual with a disability has been shown to be job-related and consistent with business necessity," and that "such performance cannot be accomplished by reasonable accommodation." At this point in time, given the speculative nature of genetic propensity for disease, employers might find it difficult to prevail on such defenses. For example, if an employer is engaging in medical testing (pre-employment, as well as monitoring and screening during employment), proving no knowledge of the "applicant's or employee's condition" might be untenable. On the other hand, as is discussed infra, statistics reveal that employers almost always prevail. Despite what appears to be at least a minimal level of protection against genetic discrimination in the workplace, the ADA has its shortcomings. First, even though the EEOC Interpretive Guidelines may be important in at least laying the foundation for bringing a case of genetic discrimination under the ADA based on an employer's perception, courts are not bound by these guidelines. The case that came the closest to testing the issue of whether an asymptomatic person is protected by the ADA is Bragdon v. Abbott. This case, however, was based on Subpart One of the ADA definition of "disability," not the "being regarded as having such an impairment" language in Subpart Three. In Bragdon, the Supreme Court held that HIV is a "disability" under the ADA, even when the infection "has not yet progressed to the so-called symptomatic phase." A key aspect of the Court's decision was its finding that because "HIV infection must be regarded as a physiological disorder with a constant and detrimental effect on the infected person's hemic and lymphatic systems from the moment of infection. HIV infection satisfies the statutory and regulatory definition of a physical impairment during every stage of the disease." Moreover, the Court agreed with respondent's claim that "the HIV infection placed a substantial limitation on her ability to reproduce and bear children" and that, for purposes of the ADA, "reproduction falls well within the phrase `major life activity.'" In reaching this determination, the Court stated that the medical evidence presented leads to the conclusion that the respondent's infection substantially limited her ability to reproduce in two independent ways. "First, a woman infected with HIV who tries to conceive a child imposes on the man a significant risk of becoming infected," and "second, an infected woman risks infecting her child during gestation and childbirth, i.e., prenatal transmission." The ability to extend Bragdon to asymptomatic genetic conditions is limited to similar circumstances in which it can be demonstrated that the condition is, indeed, a "physical or mental impairment that substantially limits one or more of the major life activities of such individual." As such, there is no reasonable sense of certainty that the ADA can be used to prevent discrimination on the basis of genetic predisposition.

Second, as Dr. Collins of the Human Genome Project has pointed out, the EEOC interpretation does not cover "unaffected recessive carriers of recessive genes" and it does not "prevent employers from obtaining a general release from an applicant once there has been a pre-employment offer." Third, and perhaps most critically, the ADA does not prevent an employer from obtaining genetic information; the law only prevents use of the information. Under the ADA, a "covered entity may require a medical examination after an offer of employment has been made and may condition an offer of employment on the results of the examination." Thus, "once the offer of employment is made, that offer may be conditioned on the outcome of a medical examination." This effectively permits an employer to learn about a potential employee's medical history, then contrive some excuse not to hire anyone who might be costly in terms of absence, productivity, insurance, etc.

Lastly, it has been exceedingly difficult for plaintiffs to prevail under the ADA. In fact, according to an American Bar Association's Commission on Mental and Physical Disability Law, in 1999, "employers prevailed in 291 cases decided in federal courts, [versus] just 13 wins for employees." That means that "employers won 95.7 percent of the time," an increase over the 1998 survey results. There is no reason to believe that this grim outlook for plaintiffs would be any different for claims based on genetic discrimination.

C. Health Insurance Portability and Accountability Act of 1996

The Health Insurance Portability and Accountability Act of 1996 (HIPAA), which amends the Employee Retirement Income Security Act (ERISA), was signed by President Clinton on August 21, 1996. HIPAA provides some federal protection against genetic discrimination in health insurance. Before HIPAA took effect, "employers and health insurers had the ability to limit or deny health insurance coverage to individuals with preexisting medical conditions." Under HIPAA, group health plans and health insurance issuers offering group health insurance coverage may impose a preexisting condition exclusion only if:

> *such exclusion relates to a condition (whether physical or mental), regardless of the cause of the condition, for which medical advice, diagnosis, care, or treatment was recommended or received within the 6-month period ending on the enrollment date; such exclusion extends for a period of not more than 12 months (or 18 months in the case of a late enrollee) after the enrollment date; and the period of any such preexisting condition exclusion is reduced by the aggregate of the periods of creditable coverage.*

The key aspect to HIPAA, with regard to genetic information, is that under the definitions, "genetic information shall not be treated as a condition described [above] in the absence of a diagnosis of the condition related to such information." This change in the law is significant, because it prohibits employers and health insurers from denying coverage to those who undergo genetic testing, unless the testing results in the "diagnosis of the condition." For example, if a woman tests positive for the gene that is linked to breast cancer, this information cannot be treated as a preexisting condition, unless an actual diagnosis of breast cancer is made within the prescribed six-month period. Moreover, if such a diagnosis/s made, under HIPAA the woman could only be excluded from coverage for a maximum of one year. Accordingly, "a group health plan, and a health insurance issuer offering group health insurance coverage in connection with a group health plan, may not establish rules for eligibility (including continued eligibility) of any individual to enroll under the terms of the plan based on" a number of factors, including medical history, genetic information, and disability. Another important aspect of HIPAA, is that it prevents insurance companies from charging one person "a premium or contribution which is greater that the premium or contribution for a similarly situated individual enrolled in the plan on the basis of any health-status related factor."

HIPAA is an important first step in addressing genetic discrimination in health insurance; however, the Act may not go far enough. First, even though individuals in group plans cannot be charged higher premiums, there is nothing in the Act to prevent insurance companies from charging all members of a group higher premiums based on the genetic testing results of one or just a few group members. The result could be to make health insurance too costly for small employers. Second, the protections pertain only to group health plans, doing nothing to protect those people in the individual market. As Senator James Jeffords pointed out during the Congressional hearings, the individual market "will be an increasingly important method of coverage" for Americans aged fifty-five to sixty-five who, with increasing frequency, are retiring early and may need to obtain insurance outside of employment. Third, HIPAA "does nothing to prevent a health insurance company from requiting that an individual submit to a genetic test as a condition of coverage, and there are no restrictions on the release of genetic testing information to others." Last, HIPAA is not comprehensive enough to "address the larger problem of gathering or using genetic information in the workplace out of the health insurance context." Thus, despite the importance of HIPAA, this legislation is merely another part in what is currently a piecemeal solution which attempts to address concerns about misappropriation of genetic information.

## D. Title VII of the Civil Rights Act of 1964

Title VII of the Civil Rights Act of 1964 (Title VII) should prohibit genetic discrimination at least in cases in which there is a disparate impact along the lines of "race, color, religion, sex, or national origin." Employers cannot "fail or refuse to hire or discharge any individual or otherwise to discriminate against any individual" because of any of these factors. Furthermore, the Pregnancy Discrimination Act provides that discrimination on the basis of "sex" includes discrimination on the "basis of pregnancy, childbirth, or related medical conditions." Inasmuch as some genetic traits seem to be associated with race, ethnicity, or sex, genetic testing may have a disproportionate effect on members of a protected class. Examples include sickle cell disease, which is associated with individuals of African descent, and Tay-Sachs disease, which is associated with Ashkenazi Jews. Thus, for instance, if an employer required pre employment health examinations for only Black and Jewish employees in search of sickle cell disease or Tay-Sachs disease, Title VII would likely be violated.

Such a situation arose in Norman-Bloodsaw v. Berkeley Laboratory. As was discussed in Part II, in this case, pre-placement tests required Black employees to undergo a test for sickle cell trait and women to undergo a test for pregnancy as a "condition of employment." Acknowledging that "the employment of women and blacks at Lawrence was conditioned in part on allegedly unconstitutional invasions of privacy to which white and/or male employees were not subjected," the Ninth Circuit upheld plaintiffs' claims as "actionable under Title VII."

Because of the complexity of interpreting the meaning of genetic information, this data will be highly amenable to misinterpretation and abuse and could be used to legitimate existing class and racial disparities on the grounds that class and race are linked to genetically-based characteristics. Should this happen, Title VII might be increasingly called upon to combat genetic discrimination in the workplace. As a general rule, however, because it appears that most gene-based diseases do not disproportionately affect individuals in the protected classes, claims for discrimination pursuant to Title VII will be limited. Moreover, proving disparate impact claims can be quite difficult. Lastly, inasmuch as redress is restricted to actions against employers who "affect" commerce and have at least fifteen employees for twenty or more weeks per year, Title VII is not adequate to address genetic discrimination in the workplace.

## E. Occupational Safety and Health Act

The Occupational Safety and Health Act (OSHA) was enacted by Congress in 1970 to "assure so far as possible every working man and woman in the Nation safe and healthful working conditions and to preserve our human resources." In an attempt to reach this ideal, OSHA imposes two duties on employers. The first, commonly known as the "general duty clause," requires employers to provide places of employment that are "free from recognized hazards that are likely to cause death or serious physical harm." This requires the employer to "eliminate significant risks of hazard as much as pcssible." The second duty, known as the "compliance clause," requires employers to comply with OSHA safety regulations and standards, Failure to maintain a workplace that meets OSHA requirements can result in civil and criminal penalties.

The requirements of OSHA can result in dilemmas for employers, especially those who are dealing with the presence of harmful workplace toxins. Biological monitoring, that is "periodic analysis of body fluids, tissues and excreta in order to measure impact of the body's exposure to chemical agents and to evaluate the health risks" that such toxins pose and workplace medical screening can be important ways of effectively detecting and preventing occupational illness. Consider, for example, the "required periodic blood tests of employees and possible removal of an employee from a high lead work area if that employee's blood lead level exceeds certain prescribed safety levels." Under such circumstances, screening can identify a worker who seems to be "susceptible to workplace exposures because the harm caused by workplace toxins generally parallels the employee's length and intensity of exposure as well as the worker's own susceptibility." Another similar situation involves OSHA regulations regarding cancer-causing agents, which provide that "before an employee is assigned to enter a regulated area, a pre assignment physical examination by a physician shall be provided. The examination shall include the personal history of the employee, family and occupational background, including genetic and environmental factors." Ironically, in a world in which fear abounds about employers conducting genetic testing and steps are being taken to limit the same, if employers fail to engage in testing pursuant to OSHA, theoretically they could run the risk of liability.

## F. United States Constitution

If the government is the employer (or if a private employer is engaged in state action or a close nexus can be established), the use of genetic screening may be unconstitutional. Depending on the circumstances of the testing, there are several possible constitutional challenges which can be asserted.

First, to the extent that individuals with a genetic mutation could comprise a suspect class, an equal protection argument can be made. Pursuant to the Fourteenth Amendment, states are prohibited from depriving any person equal protection of the laws. Similar to the rationale for bringing a claim for genetic discrimination under Title VII, plaintiffs who fall within a suspect class (such as race and sex) may find recourse under an equal protection claim. Moreover, it may be possible for individuals who have genetic markers identifying them as at-risk for or carders of a genetic condition to contend that they should be defined as a "suspect class."

Second, to the extent that a right to privacy has been read into the U.S. Constitution since the Supreme Court's decision in Griswald v. Connecticut, an argument could be made that an individual's genetic profile should be protected against disclosure. Although a right to privacy is not explicitly set forth in the Constitution, the Supreme Court read the First, Third, Fourth, and Fifth Amendments to recognize such a penumbral right. In Norman-Bloodsaw v. Lawrence Berkeley Laboratory, the Ninth Circuit stated that the "constitutionally protected privacy interest in avoiding disclosure of personal matters clearly encompasses medical information and its confidentiality." Acknowledging that cases "defining the privacy interest in medical information have typically involved its disclosure to `third' parties, rather than the collection of information by illicit means, it goes without saying that the most basic violation possible involves the performance of unauthorized tests, that is, the non-consensual retrieval of previously unrevealed medical information that may be unknown even to plaintiffs." Norman-Bloodsaw clearly lays the groundwork to challenge the government's right to obtain genetic information.

Third, and a point related to the right to privacy, unauthorized tests that reveal personal medical information may also be viewed as searches in violation of the Fourth Amendment, 177 which protects people against unreasonable searches and seizures. Fourth Amendment scrutiny requires balancing the government's interest in conducting these particular tests against plaintiffs' expectations of privacy.

Thus, there are several challenges to the unauthorized gathering of genetic information that can be asserted under the U.S. Constitution. Yet, similar to the other possible avenues of recourse under federal law, constitutional challenges are limited in their scope. Most workers are left without adequate protection from genetic discrimination and employers are left without a clear sense of what degree (if any) of genetic testing and monitoring is defensible.

## Existing Protection against the Discriminatory use of Genetic Information in the Workplace: State Laws

### A. Statutory Law

Many states have passed laws protecting workers from genetic discrimination in the workplace. The breadth of worker protection ranges from no protection to comprehensive mandates against the use of genetic information in the employment context. In general, state laws prohibiting genetic discrimination can be categorized into three broad types: laws that prohibit discrimination based on the trait for a specific disease; laws that prohibit discrimination based upon genetic test results (broadly termed "genetic status"); and laws that prohibit discrimination based on genetic test results and also prohibit testing to ascertain that status.

Examples of laws that prohibit employers from discriminating against individuals possessing particular genetic traits or disorders can be seen in Florida (sickle cell trait), Louisiana (sickle cell trait), North Carolina (sickle cell trait and hemoglobin C trait), New Jersey (sickle cell trait, hemoglobin C trait, thalassemia trait, Tay-Sachs trait, and Cystic Fibrosis trait), and New York (sickle cell trait, Tay-Sachs trait, and [beta]-thalassemia trait). The second category, laws that prohibit discrimination in the workplace based on genetic status, are often criticized because they lack a provision regarding genetic testing. In Arizona, employers "may not fail or refuse to hire, discharge, or discriminate on the basis of genetic test results." Connecticut and North Carolina have similar laws prohibiting genetic discrimination, and which do not address the issue of testing. The approach taken in Illinois is to require employers to "treat genetic testing information in a manner consistent with federal law, including the ADA," and provides that, except under limited circumstances, no one "may disclose or be forced to disclose the identity of someone who has been genetically tested or to disclose his genetic information." Unlike some of the more comprehensive statutes, such as New Jersey's, the law in these states is very brief and limited in its application.

The third category of laws prohibits workplace discrimination based on genetic information and also restricts the gathering of genetic information. Each state varies in its level of protection. New

Jersey's Genetic Privacy Act, however, is often cited for the expansive degree of protection it provides workers in that state. The law explicitly acknowledges that the "improper collection, retention or disclosure of genetic information can lead to significant harm to the individual, including stigmatization and discrimination in areas such as employment, education, health care and insurance." Accordingly, in New Jersey, it is "an unlawful employment practice for an employer to refuse to hire, to discharge, to force to retire an employee based on genetic information or to discriminate in compensation or terms, conditions or privileges of employment." Moreover, informed consent is needed to obtain, retain, or disclose genetic information (with certain exceptions not related to employment) and "notice shall be provided to the genetic test subject that the test was performed and results were received." The disclosure of genetic information is then restricted, prohibiting the disclosure of the identity of the person upon whom the test was performed. Another key feature of the New Jersey legislation is the provision which provides fines, imprisonment, and actual damages for violations.

New York also has more defined aspects to its laws prohibiting genetic discrimination. Specifically with regard to testing, prior and specific written informed consent is required to perform a genetic test, and prior written informed consent is required for the disclosure of genetic test results. New York law also specifically prohibits employers, licensing agents, employment agencies, and labor organizations from soliciting, requiring, or administering a genetic test or buying or acquiring genetic test results. Notwithstanding this provision, such entities may require a specific genetic test when it is directly related to the occupational environment and certain employees may be genetically predisposed to an increased risk; even under these circumstances, however, the employee must give written consent.

As is apparent from this survey of statutes, the degree of protection for genetic discrimination in the workplace varies significantly from state to state. Such variations include: the definitions for "genetic testing," "genetic information," "genetic predisposition," and "genetic characteristic;" applicability to insurance coverage; the circumstances surrounding testing (including consent and when an employer has a right to test); the maintenance of records related to genetic testing; the determination of when disclosure of a genetic condition is necessary; and the penalties for improper use and disclosure of genetic information.

## B. State Constitutions

Under claims similar to those discussed above regarding the U.S. Constitution, claims for genetic discrimination may also be made pursuant to some state constitutions. In California, for example, the state constitution includes a right to privacy. To assert a cause of action under Article I, Section 1 of the California Constitution, "one must establish three elements: a legally protected privacy interest; a reasonable expectation of privacy under the circumstances; and conduct by the defendant that amounts to a 'serious invasion' of the protected privacy interest." As is the case for state legislation, protection for genetic discrimination under state constitutions varies by state and, accordingly, lacks any kind of uniformity. The overall deficiency of any kind of uniform state protection underscores the need for federal legislation.

## Genetic Nondiscrimination in Health Insurance and Employment Act of 1999

The most significant and comprehensive pending legislation. designed to address genetic discrimination is the Genetic Nondiscrimination in Health Insurance and Employment Act of 1999 (the Act). The bill sponsors, Senator Thomas Daschle and Representative Louise Slaughter, have embarked on a media campaign to raise awareness about the need for this legislation. On June 26, 2000, the day that the announcement was made about the status of the Human Genome Project, Slaughter maintained that "social policy must keep pace with science. The abuse of genetic information must be prevented." Despite the tremendous promise of this science, Slaughter warned that "many Americans will refuse to take genetic tests unless they are assured their information will remain private. Each of us has between five and 50 faulty genes. We want to make sure health insurance companies and employers won't drop people because they might, or might not, get sick at some point in the future." Two days later, Daschle was quick to concur, stating that "we have to ensure that our laws keep pace with these amazing advances in science and technology. Many Americans are increasingly afraid that their genetic information may be used against them and will jeopardize their ability to get a job or obtain insurance."

Most recently, in an attempt to bring the Genetic Nondiscrimination in Health Insurance and Employment Act to a vote in the House of Representatives, Slaughter launched a "discharge petition." Under this procedural move, the full House of Representatives will consider a bill of a majority if the 218 members of the House sign the petition. This legislation, which remains pending in the House and Senate, would eradicate the significant differences created by state genetic discrimination legislation by: creating uniform definitions that ensure equal treatment of individuals and their families no matter where they reside; providing for a clear directive for handling genetic information in the context of both group and individual health insurance plans; specifying limitations on genetic testing; delineating limitations on genetic testing; setting forth rules regarding the collection and disclosure of predictive genetic information; furnishing uniform guidelines prohibiting employment discrimination on the basis of predictive genetic information; and establishing penalties for violations.

**A) Title I: Amendments to ERISA, PHSA, IRC, and Title XVIII of the SSA**

With respect to genetic information, Title I of the Genetic Nondiscrimination in Health Insurance and Employment Act is designed to amend the Employee Retirement Income Security Act of 1974 (ERISA), regarding a group health plan and a health insurance issuer offering group insurance through a group plan; the Public Health Service Act (PHSA), regarding both group health plans and the individual market; the Internal Revenue Code, regarding a group health plan; and Title XVIII (Medicare) of the Social Security Act, regarding Medicare supplemental policies. This is sweeping legislation intended to have far-reaching ramifications in a number of realms in which genetic discrimination could occur. Each of these laws would be amended to add definitions for "family member," "genetic information," "genetic services," "genetic test," and "predictive genetic information," to clearly set forth what individuals and medical information are covered.

These definitions are comprehensive and much more detailed than the current legislation in many states. "Family" is broadly construed, "genetic services" encompasses both interpretation of genetic information and any counseling, and "genetic test" extends beyond human DNA to also include RNA, chromosomes, proteins, and certain metabolites. The intent behind this legislation is to anticipate, as best one can, technological changes in genetic research and to extend protection accordingly.

Moreover, ERISA, the PHSA, the IRC and the SSA would be amended to prohibit discriminating in individual enrollment, including adjusting insurance rates on the basis of predictive genetic information; discriminating in group eligibility or group premium or contribution rates; requesting or requiring test performance; and requesting, requiting, collecting, purchasing, or disclosing information, unless authorized by the individual.

Additionally, the legislation would allow a group health plan or health insurance issuer with regard to payment for genetic services, to "request that the individual provide the plan or issuer with evidence that services were performed." With respect to "the payment of claims for benefits other than for genetic services," predictive genetic information may be required from an individual, so long as such information " is used solely for the payment of a claim; is limited to information that is directly related to and necessary for the payment of such claim and the claim would otherwise be denied but for the predictive genetic information; and is used only by an individual (or individuals) within such plan or issuer who needs access to such information for purposes of payment of a claim."

Should violations of Title I of.the Act occur, the court is given the tight to award any appropriate legal and equitable relief, which may include "a requirement for the payment of attorney's fees and costs, including the costs of an expert witness."

**B) Title II: Prohibition of Employment Discrimination on the Basis of Predictive Genetic Information** Title II of the Genetic Nondiscrimination in Health Insurance and Employment Act of 1999 prohibits employment discrimination on the basis of predictive genetic information. Under this legislation, it shall be an unlawful employment practice for an employer "to fail or refuse to hire or to discharge any individual, or otherwise to discriminate against any individual with respect to the compensation, terms, conditions, or privileges of employment of the individual, because of predictive genetic information with respect to the individual (or information about a request for or the receipt of genetic services by such individual or family member of such individual)." The same definitions from Title I apply to Title II, affording individuals and their family members broad protection against genetic discrimination in the workplace. Limiting, segregating, or classifying employees of the employer in any way that "would deprive or tend to deprive any individual of employment opportunities or otherwise adversely affect the status of the individual as an employee, because of predictive genetic information" is also prohibited. The Act limits the ability of an employer to "request, require, collect or purchase predictive genetic information" with respect to an individual or a family member, except under limited circumstances. In any event, any information learned under these exceptions may not be used in violation of the Act. This section is designed to balance the privacy of the worker with any need of the employer to know genetic information to be in compliance with OSHA and the MSHA. An important aspect of the genetic monitoring, however, is the requirement that workers should be fully informed about the testing and results.

Similar to the provisions that apply to employers, employment agencies and labor organizations are prohibited from taking any action that discriminates against an individual because of predictive genetic information, including failing or refusing to refer an individual for employment and excluding or expelling a member, respectively. It is also considered an "unlawful employment practice" for both employment agencies and labor organizations to limit, segregate, or classify individuals/members or fail or refuse to refer for employment any individual in a way that would deprive or tend to deprive any individual of employment opportunities or would limit the employment opportunities or otherwise adversely affect the status of the individual as an employee, because of predictive genetic information with respect to the individual (or information about a request for or the receipt of genetic services by such individual or family member of such individual); or to request, require, collect, or purchase predictive genetic information with respect to an individual (or information about a request for or the receipt of genetic services by such individual or family member of such individual); or to cause or attempt to cause an employer to discriminate against an individual in violation of this title. Likewise, similar provisions are set forth governing any "employer, labor organization, or joint labor-management committee controlling apprenticeship or other training or retraining, including on-the-job training programs."

A very important aspect of the Act is the provision for the maintenance and disclosure of predictive genetic information. The Act safeguards genetic information by requiring all employers who possess "predictive genetic information about an employee (or information about a request for or receipt of genetic services by such employee of family member of such employee)" to treat or maintain such information "as part of the employee's confidential medical records." Disclosure of predictive medical information is prohibited, unless the employee who is the subject of the information requests disclosure; the information is requested by an occupational or other health researcher (if the research is conducted in compliance with the regulations and protections provided for under Part 46 of Title 45, Code of Federal Regulations); there is legal compulsion of a federal court order (with restrictions); and government officials investigating compliance with this Act request the information as relevant to the investigation.

Enforcement power is given to the EEOC. Aggrieved employees, members of labor unions, and participants in training programs are given the right to bring a civil action in federal or state court against an employer, employment agency, labor organization, joint labor-management committee, or training program who violate the Act. Moreover, nothing in Title II is to be construed to limit the rights and protections under the ADA, the Rehabilitation Act of 1973, or any other federal or state statute that provides equal or greater protection to an individual. The Act does not apply to the Armed Forces Repository of Specimen Samples for the Identification of Remains, and does not limit the statutory or regulatory authority of OSHA or the MSHA "to promulgate or enforce workplace safety and health laws and regulations."

Overall, the Act offers a substantial level of protection for employees while, at the same time, allowing employers access to selected information necessary to maintain a safe work environment.

## Conclusion

The calls for increased protection from genetic discrimination and for refined definitions of when the use of genetic information should be permitted are largely addressed by the proposed Genetic Nondiscrimination in Health Insurance and Employment Act of 1999. This Act would alleviate a good bit of the fear that genetic information will be misused, thereby allowing science to advance and providing incentive to individuals to reap the benefits of genetic research. The Act would create comprehensive uniform federal protection, which would go a long way toward ensuring that all Americans have access to medical care and information which can help them live longer and healthier lives.

Yet, despite this fact, not everyone is enthusiastic about legislating to prevent genetic discrimination. Somewhat surprisingly, Michael Kinsley, formerly the voice on the "Left" on CNN's Crossfire, claims that "genetic discrimination is universal, inevitable and, in some ways, essential." He argues that "discrimination makes perfect sense" for health insurers who want to keep people out of their "insurance pool if they're more likely to get sick. The idea of insurance being to protect against unpredictable costs." When Burlington Northern Santa Fe Railroad initiated tests on employees who claim to have work-related carpal-tunnel syndrome to determine if they are genetically susceptible to this repetitive-stress injury, they were likely trying to cut costs. After all, if the company could demonstrate that the injuries were due to a biological predisposition and had nothing to do with their employment, Burlington Northern could avoid paying claims. Likewise, the disclosure that Norwich Union Life, one of Britain's largest insurance companies, was forced to admit that it was using unapproved genetic tests for potentially fatal diseases when deciding whether to underwrite life insurance applications comes as no surprise. As genetic information becomes available, employers and other entities want to make full use of the information to cut costs.

Social commentator Andrew Sullivan further underscores this point by using an analogy to racial discrimination. He says that the "point of laws against racial bias is to outlaw irrational discrimination based on irrelevant characteristics. The point of laws against genetic discrimination is to outlaw rational bias based on relevant information. The two principles aren't merely different; they're opposite." Sullivan concludes his editorial with the foreboding prophesy that "men and women will increasingly be judged not by the color of their skin but by the content of their chromosomes. The irrational cruelty of bias will be replaced by the rational cruelty of fate." Kinsley and Sullivan have a point. Once the proverbial genetic genie is out of the bottle, there is virtually no way to categorically deny access to such information. The challenge becomes how to fairly utilize information about genetic predisposition without creating a situation in which the new resume will be merely a DNA printout. Such is the thematic basis of the 1997 film Gattaca in which those who are permitted to gain entry into the most prestigious positions are genetically engineered to be superior. In this bleak futuristic society, genetic essentialism creates a new underclass of the genetically inferior. These so-called "invalids" are presumed to be incapable of rising above any genetic defects and are therefore relegated to the most menial labor. The love-child (i.e., unscientifically conceived or altered) hero of the film engages in biological deception to make the point that sheer vision, will, and determination can compensate for his grim genetic destiny. Ultimately, he prevails, defying those who could not see past his DNA.

In recent years our nation's courts have come to accept DNA evidence as absolute proof of innocence. For example, genetic evidence has been used to free death-row inmates and to exonerate individuals wrongly convicted of rape and other crimes. Even so, courts have been slow to accept DNA as proof when a man contends he should not have to pay child support for a child that is not his.

State legislators have sought to ease the financial burden of welfare payments by demanding that aid recipients name the fathers of their children. Laws that once sought to distribute welfare payments by collecting money from former spouses and lovers have been manipulated into tools for extorting child-support payments from men who did not father the children.

For centuries, courts followed a rule known as the "presumption of paternity." This tenet of common law states that unless a man can prove that he is sterile, impotent, or was away from home at the time of conception, he is the legal father of any child born to his wife during their marriage. The Romans first adopted this rule, and the English incorporated it into common law some 500 years ago.

Despite scientific advances and case law that supports the use of DNA tests as evidence in exonerating the accused, the presumption of paternity remains in practice even today. Because of this, courts have allowed mothers to commit perjury, assigning paternity to former spouses or, in the cases of those never married, former lovers.

Because none of the 50 states requires a mother who files a claim for child support to advise the court or child-support agencies when another man potentially could be the father, thousands of men are paying support for children who may not be their own. This deception has come to be known in legal circles as "paternity fraud."

Many states have opted to look the other way when it comes to adopting legislation against this type of deception. Only two states have instituted legislation that allows men unlimited time to challenge paternity using DNA testing: Maryland, which passed legislation in 1995, and Ohio, which passed its bill in 2000. The Georgia Legislature recently passed a paternity-fraud bill that now awaits the governor's signature.

Several states, including California and Colorado, have similar legislative proposals that would address this growing problem. The California Paternity Justice Act of 2002 would require DNA testing in cases of disputed paternity. Other states have addressed this problem by limiting paternity challenges: Iowa allows a maximum of three years for such challenges, Colorado allows five years and Louisiana 10 years.

Many of the men assigned default judgments for child support never received summonses to appear in court. Because of the prevalence of default judgments, men who have not fathered a child can turn into "dads" instantly when the court enters a default judgment.

Once applied, that label is difficult, if not impossible, to remove. State agencies and district attorneys frequently fight such appeals, challenging anyone who fails to respond to a court summons even one served to a "last-known" address. This aggressive pursuit of assigning paternity at any cost has ensnared thousands of innocent men mistakenly identified as having fathered children who do not belong to them.

The prevalence of paternity fraud has reached startling proportions. As many as 28 to 30 percent of men tested for paternity learn they are not biologically related to the children they allegedly fathered, according to the American Association of Blood Banks.

The high incidence of default paternity judgments can be tied to economic motives: More than 11 million fathers do not live with their children. Nationwide, child-support collections have grown to more than $18 billion annually. State legislatures long have sought to offset the financial drain of welfare payments made to single mothers by identifying biological fathers who do not live with their children.

State child-support agencies receive federal reimbursements of two-thirds of their administrative budgets, in addition to incentives of 6 to 10 percent for each dollar of child support collected. Under federal guidelines, states must identify the fathers of children whose mothers are receiving welfare benefits or risk losing incentive monies. Federal guidelines also encourage aggressive collection efforts by allowing states to retain the cost of welfare benefits from child-support payments made by the non custodial parent. Because federal rules do not require DNA testing to prove paternity, states have no incentive to investigate claims of fraud and every reason to fight efforts to disestablish paternity.

Private child-support agencies have similar motives in refusing to acknowledge DNA evidence disproving paternity. Mothers routinely contact private agencies in an attempt to collect past-due court-ordered child support. In enforcing court orders, these agencies routinely collect 30 percent of the past-due amount, in addition to 30 percent of all future payments. Because these agencies enforce court orders, they have no reason to cease collection efforts once presented with DNA evidence refuting paternity.

The true impact of paternity fraud goes far beyond the dollars collected from men who are not fathers, however. For children facing life-threatening illnesses, such as cancers requiring bone-marrow transplants or other medical emergencies, knowing one's biological heritage can be a matter of life or death. While courts acknowledge and rightfully respect the confidentiality of adoptive parents, the practice of assigning paternity to uninformed and unrelated individuals goes beyond fraud it potentially places the lives of children in danger. Some argue that demanding accuracy in determinations of fatherhood would harm children who have established relationships with men they believe to be their fathers. Make no mistake: In most cases of paternity fraud, the children have never even met the supposed father. Furthermore, court proceedings for assigning fatherhood do not seek to establish a relationship between a parent and a child. Instead, paternity hearings seek to establish only a financial obligation on the part of the alleged father to the child.

Because states refuse to prosecute mothers who make false claims about their child's paternity, countless men are forced to spend thousands of dollars to clear their names while also supporting children fathered by others. Consider the following cases:

Interestingly, women from a variety of financial backgrounds provide fraudulent information regarding their children's paternity. Who they name as the father oftentimes depends on their marital status. Married women most often name the men who were their husbands at the time of conception, fulfilling a societal expectation of monogamy. However, unmarried women who have multiple intimate partners most often name the men who will best provide financially for their children. In some cases, mothers will implicate men whom they believe will fill the role of a father to their children.

When courts allow mothers to pick and choose who will be financially responsible for their children, regardless of tree parentage, fraud will prevail. When the troth no longer matters, state agencies tend to decide they have no obligation to seek out biological fathers and refuse to acknowledge DNA evidence disproving paternity.

# Bibliography

Altschul, S. F., and D. J. Lipman. ( 1989) "Trees, stars, and multiple biological sequence alignment". *SIAM J. Appl. Math. 49*:197-209.

Amorese, D. A., and A. M. Hochberg. ( 1989) "A fluorescence-labeled DNA analysis system for sequencing and mapping". *Am. Biotech. Lab. 7*:38-46.

Arratia, R., and M. S. Waterman. ( 1989) The Erdös-Rényi strong law for pattern matching with a given proportion of mismatches. *Annals Prob. 17*:11521169.

Arratia, R., L. Gordon, and M. S. Waterman. ( 1986) An extreme value theory for sequence matching. *Ann. Statist. 14*:971-993.

Bell, G. I., and T. G. Marr . (eds.). ( 1989) *Computers and DNA: The Proceedings of the Interface Between Computation Science and Nucleic Acid Sequencing Workshop*. Santa Fe Institute Studies in the Series of Complexity, vol. 7. Addison-Wesley, Redwood City.

Carrillo, H., and D. Lipman. ( 1988) The multiple sequence alignment problem in biology. *SIAM J. Appl. Math. 48*:1073-1082.

Chen, E. J., and P. H. Seeburg. ( 1985) "Supercoil sequencing: a fast and simple method for sequencing plasmid DNA". *DNA* 4:165-170.

Dayhoff, M. O. (ed.). ( 1979) *Atlas of Protein Sequence and Structure*, vol. 5, suppl. 3. National Biomedical Research Foundation, Washington, D.C.

Doolittle, R. F. (ed.). ( 1990) *Molecular Evolution: Computer Analysis of Protein and Nucleic Acid Sequences*. Methods in Enzymology, vol. 183. Academic Press, New York.

Erdmann, V. A., E. Huysman, A. Vandenberghe, and R. De Wachter. ( 1983) Collection of published 5S and 5.8S ribosomal RNA sequences. *Nucleic Acids Res. 11*:r105-r133.

Erdös, P., and A. Rényi. ( 1970) On a new law of large numbers. *J. Anal. Math. 22*:103-111; reprinted in "Selected Papers of Alfred Rényi", vol. 3, 1962- 1970. Akademiai Kiado, Budapest, 1976.

Erlich, H. A. (ed.). ( 1989) *PCR Technology. Principles and Applications for DNA Amplification*. Stockton Press, New York.

Feng, D., and R. F. Doolittle. ( 1987) Progressive sequence alignment as a prerequisite to correct phylogenetic trees. *J. Mol. Evol. 25*:351-360.

Fitch, W. M. ( 1971). Towards defining the course of evolution: minimum change for a specific tree topology. *Syst. Zool. 20*:406-416.

Fitch, W. M., and E. Margoliash. ( 1967) Construction of phylogenetic trees. *Science 155*:279-284.

Galas, D. J., M. Eggert, and M. S. Waterman. ( 1985) Rigorous pattern recognition methods for DNA sequences: analysis of promoter sequences from *E. coli. J. Mol. Biol. 186*:117-128.

Gotoh, O. ( 1982) An improved algorithm for matching biological sequences. *J. Mol. Biol. 162*:705-708.

Hein, J. ( 1989) A new method that simultaneously aligns and reconstructs ancestral sequences for any number of homologous sequences, when the phylogeny is given. *Mol. Biol. Evol. 6*:649-668.

Johnson, M. S., and R. F. Doolittle. ( 1986) A method for the simultaneous alignment of three or more amino acid sequences. *J. Mol. Evol. 23*:267-278.

Kohara, Y., K. Akiyama, and K. Isono. ( 1987) The physical map of the whole E. coli chromosome: application of a new strategy for rapid analysis and sorting of a large genomic library. *Cell 50*:495-508.

Lipman, D. J., and W. R. Pearson. ( 1985) Rapid and sensitive protein similarity searches. *Science 227*:1435-1441.

Lipman, D. J., S. F. Altschul, and J. D. Keceioglu. ( 1989) A tool for multiple sequence alignment. *Proc. Natl. Acad. Sci. USA 86*:4412-4415.

Maxam, A., and W. Gilbert. ( 1977) "A new method for sequencing DNA". Proc. Natl. Acad. Sci. USA 74:560-564.

Maxam, A., and W. Gilbert. ( 1980) "Sequencing end-labeled DNA with base specific chemical cleavage. Methods Enzymol". 65:499-560.

Morelle, G. ( 1989) "A plasmid extraction procedure on a miniprep scale". BRL Focus 11:7-8.

Naim, C. J., M. Miyamoto, and R. J. Ferl. ( 1991) Phylogenetic implication of the DNA sequence of the 18s rRNA gene of Gingko biloba. *Mol. Biol. Evol.*, submitted.

Needleman, S. B., and C. D. Wunsch. ( 1970) A general method applicable to the search for similarities in the amino acid sequences of two proteins. *J. Mol. Biol. 48*:444-453.

Pearson, W.R, and D. J. Lipman. (1988) Improved tools for biological sequence comparison. *Proc. Natl. Acad. Sci. USA 85*:2444-2448.

Prober, J. M., G. L. Trainor, R. J. Dam, F. W. Hobbs, C. W. Robertson, R. J. Zagursky , A. J. Cocuzza, M. A. Jensen, and K. Baumeister. ( 1987) "A system for rapid DNA sequencing with fluorescent chain-terminating dideoxynucleotides". *Science 238*:336-341.

Sanger, F., S. Nicklen, and A. R. Coulson. ( 1977) "DNA sequencing with chain terminating inhibitors". *Proc Natl. Acad. Sci. USA* 74:5463-5467.

Sankoff, D. ( 1975) Minimal mutation trees of sequences. *SIAM J. Appl. Math. 28*:35-42.

Sellers, P. ( 1974) On the theory and computation of evolutionary distances. *SIAM J. Appl. Math. 26*:787-793.

She, J. X., F. Bonhornme, E. Desmarais, and E. K. Wakeland. ( 1991) Mitochondrial DNA sequence evolution and phylogeny in the genus Mus. *Mol. Biol. Evol.*, submitted.

Smith, L. M., J. Z. Sanders, R. J. Kaiser, P. Hughes, C. Dodd, C. R. Connell, C. Heiner , S. B.H. Kent, and L. Hood. ( 1986) "Fluorescence detection in automated DNA sequence analysis". *Nature 321*:674-679.

Smith, T. F., and M. S. Waterman. ( 1981) Identification of common molecular subsequences. *J. Mol. BioL 147*:195-197.

Smith, T. F., M. S. Waterman, and C. Burks. ( 1985) The statistical distribution of nucleic acid similarities. *Nucleic Acids Res. 13*:645-656.

Smith, T. F., M. S. Waterman, and W. M. Fitch. ( 1981) Comparative biosequence metrics. *J. Mol. Evol. 18*:38-46.

Tavaré, S. ( 1986) Some probabilistic and statistical problems in the analysis of DNA sequences. *Lec. Math. Life. Sci., Am. Math. Soc. 17*:57-86.

Taylor, W. R. ( 1986) Identification of protein sequence homology by consensus template alignment. *J. Mol. Biol. 188*:233-258.

Taylor, W. R. ( 1987) Multiple sequence alignment by a pairwise algorithm. *CABIOS* 3:81-87.

Waterman, M. S. ( 1986) Multiple sequence alignment by consensus. *Nucleic Acids Res. 14*:9095-9102.

Waterman, M. S., and M. Eggert. ( 1987) A new algorithm for best subsequence alignments with application to tRNA-rRNA comparisons. *J. Mol. Biol. 197*:723-728.

Waterman, M. S., and R. Jones. ( 1990) "Consensus methods for DNA and protein sequence alignment". Pp. 221-237 in *Molecular Evolution: Computer Analysis of Protein and Nucleic Acid Sequences* ( R. F. Doolittle, ed.). *Methods in Enzymology*, vol. 183. Academic Press, New York.

Waterman, M. S., L. Gordon, and R. Arratia. ( 1987) Phase transitions in sequence matches and nucleic acid structure. *Proc. Natl. Acad. Sci. USA 84*:12391243.

Waterman, M. S., T. F. Smith, and W. A. Beyer. ( 1976) Some biological sequence metrics. *Advances in Math. 20*:367-387.

Wilber, W. J., and D. J. Lipman. ( 1983) Rapid similarity searches of nucleic acid and protein data banks. *Proc. Natl. Acad. Sci. USA 80*:726-730.

# Index


A

Adenine, 149
Agent Presley, 56
Alignment Score, 97
American Association for the Advancement of Science (AAAS), 200
Amino Acid Replacements, 23
Anomalous Provisions, 143
Attorney General Janet Reno, 109

B

Barbary Macaque (M. Sylvanus), 15
Bootstrap Analysis, 24
Bronx Criminal Courts, 46
Budowle's Leadership, 58
Bureau of Justice Statistics, 123

C

Canadian Centre for Forensic Sciences, 179
Capillary Electrophoresis (CE), 4
Cases, 8
Central Park Jogger Case, 61
Chinky, 40
Cold Hit, 111
  Spring Harbor Laboratory, 46
Combined DNA Index System (CODIS), 108
Common Chimpanzee (Pan Troglodytes), 15
Compulsory Process, 176
  Process Clause, 179
Confrontation Clause, 178
Cost-Effectiveness, 127
Crab-Eating Macaque (M. Fascicularis), 15
Criminal Justice System, 92
Current Discovery Regime, 169
Cytosine, 126

D

D.H. Kaye, 74
Data, 15
Developmental Validation Studies, 167
District Attorney's Office, 44

DNA Computing, 96
- Databases, 105
- Databases for 21st Century, 147
- Effectiveness: Prosecution, 120
- Fingerprinting Methods, 2
- Laboratories, 128
- Matching, 62
- Mining, 115
- Mining Process, 112
- Molecule, 149
- Profile, 63
- Profile Match Probability Calculation, 80
- Technology in Forensic Science, 74
- Testing, 1
- Typing, 107
- War, 63

Double Helix, 46
Due Process Clause, 180
Dynamic Programming Alignment, 99

E

Economy of Scale, 130
Equal Employment Opportunity Commission (EEOC), 201
Evidentiary Reliability, 76
Evolutionary Rate, 29

F

Fake DNA Evidence, 8
FBI, 57
FBI's DNA-Testing Program, 93
Federal Convicted Offender DNA Database, 129
- Justice Statistics Database, 130

Fingerprint, 89
Forensic DNA Kits, 162
Free Market, 71
- Proof, 71

Function Creep, 135

G

Genetic Fingerprinting, 1
- Information, 214
- Monitoring, 192
- Predisposition, 214
- Screening, 193
- Status, 213
- Testing, 214

Gibbon (Hylobates Lar), 15
Global Parsimony Analysis, 20
- Bootstrap, 25

Gorilla (Gorilla Gorilla), 15
Guanine, 126

H

Human (Homo Sapiens), 15
- Genome Initiative, 96

I

Inher Inconsistent, 73

J

Japanese Macaque (Macaca Fuscata), 15
Journal of Molecular Evolution and Molecular Biology And Evolution, 96
Junk Science, 53, 94

L

Leading Molecular Biologists, 54
Life Codes's, 51

M

Macarthur "Genius" Fellowship, 47
Market Failure, 72
Matching and Binning DNA Fragments in Fore Science, 76
Matching, 63
Michigan School, 74
Misrepresentation, 54
Mitochondrial DNA (MTDNA), 6
Multi-Locus Genot, 63
  Probe, 58
Multiple Sequence Alignment, 103

N

National DNA Database (NDNAD), 2
  DNA Index System (NDIS), 109
  Institute of Justice, 109
  Opinion Research Center (NORC), 193
Non-Index Crimes, 116
Null Hypothesis, 77

O

Occupational Safety and Health Act (OSHA), 210
Offender DNA Databases, 151
Orangutan (Pongo Pygmaeus), 15

P

Philippine Tarsier (Tarsius Syrichta), 15
Phylogenetic Analysis Using Parsimony (PAUP), 15
Phylogenetic Signal, 29
Polymerase Chain Reaction (PCR), 3
Population Structure, 62
Power of DNA, 85
Probative Value, 65
Professor Thompson, 70
Protected Genetic Information, 201
Public Disservice, 94

Q

Quantifying Probative Value, 78

R

Red Herrings, 94
Relevant Evidence, 73
Reliability and Informativeness, 14
  of Methods, 14
Restriction Enzyme, 75
Restriction Fragment Length Polymorphism (RFLP), 2, 108
Rhesus Macaque (M. Mulatta), 15
Ribosomal Rna, 97
Ring-Tailed Lemur (Lemur Caua), 15
Rivera, 40

S

Scheck and Neufeld, 52
Sequence Alignment, 97
Short Tandem Repeats (STR), 4, 108
Similarity Likelihood Ratio, 66
Single Locus VNTR Probes, 63

Nucleotide Polymorphisms (SNPs), 3
Locus Probe, 58
Sir Alec Jeffreys, 1
Smycoplasma Capricolum, 101
Squirrel Monkey (Saimiri Sciureus), 15
Susceptibility to Workplace Hazards, 195

**T**

Technical Working Group for DNA Analysis Methods (TWGDAM), 168
Texas Department of Public Safety, 110
DNA Database System, 137
The Data, 15
Theory of Evidence, 75
Thompson's Argument, 68
Thymine, 126
Trade Secret/Innovation Argument, 172
Transversion Parsimony Analysis, 20
Bootstrap, 25
Trigger Mechanisms, 132
True Phylogeny, 34

**U**

University of Michigan Law School, 74

**V**

Variable Number of Tandem Repeats, 149
Number Tandem Repeat (VNTR), 4
Vilma Ponce's, 52
Violent Index Crimes, 113
Virginia Department of Corrections, 155

**W**

Washington Supreme Court, 158
Weight, 166

**X**

X-Ray Images, 3

**Y**

Y-Chromosome (Y-STR), 6